Land Acquisition in Asia

Naoyuki Yoshino • Saumik Paul
Editors

Land Acquisition in Asia

Towards a Sustainable Policy Framework

palgrave
macmillan

Editors
Naoyuki Yoshino
Asian Development Bank
Institute (ADBI)
Tokyo, Japan

Saumik Paul
Asian Development Bank
Institute (ADBI)
Tokyo, Japan

ISBN 978-981-13-6457-0 ISBN 978-981-13-6455-6 (eBook)
https://doi.org/10.1007/978-981-13-6455-6

This Palgrave Macmillan imprint is published by the registered company Springer Nature Singapore Pte Ltd.
The registered company address is: 152 Beach Road, #21-01/04 Gateway East, Singapore 189721, Singapore

Contents

Notes on Contributors

Shreyas P. Bharule is a doctoral candidate at the University of Tokyo, a member of City Planning Institute of Japan (CPIJ), while concurrently working as a research associate at Asian Development Bank Institute (ADBI), Tokyo. He completed his bachelor's in architecture at Sir. J.J. college of Architecture, Mumbai, and his masters in city planning from IIT Kharagpur.

Sayak Khatua is a research associate at the International Initiative for Impact Evaluation (3ie), New Delhi. At 3ie, he is responsible for managing research transparency initiatives; for curation, analysis, and verification of 3ie-funded impact evaluation data; and for training researchers on good data management practices. His research interests lie in the field of migration and inequality. He is an economist by training with a master's degree from Portland State University.

Saloni Lakhia is a Rotary Peace Fellow pursuing her master's in peace under the Public Policy and Social Research Program at International Christian University, Tokyo, while working part time as a research associate with the Asian Development Bank Institute, Tokyo.

She holds a bachelor's in law and has provided services to nonprofit and state government institutions in India, focusing predominantly on policy measures and legal reforms. Her research interests focus on public policy, development, and the intersection of government services with implementation imperatives.

Lai Ming Lam is Associate Professor of Anthropology at Osaka University. She holds a PhD in anthropology and master's degree in environmental studies from the University of Adelaide, Australia. She has been carrying out research on the welfare impact of conservation-led displacement on local communities, particularly in Nepal, since 2002. Over the past 11 years she had dedicated her life to improving the livelihoods of rural populations, and she is the founder of Future Village Nepal. When the 2015 earthquake hit rural Nepal, her NGO not only provided emergency relief but also implemented long-term disaster recovery and reconstruction projects. In her research projects, she examines a variety issues in regard to post-2015 Nepal earthquake, which include internal displacement and resettlement, environmental and tourism impacts, and changes in the state-civil society relationship.

Yoko Oishi is a research associate at Asian Development Bank Institute (ADBI). Before joining ADBI, she worked at Food Agriculture Organization of the United Nations as a project intern. Oishi graduated from International Christian University with a bachelor's in liberal arts and majored in business administration. She holds a master's degree and PhD in public economics from National Graduate Institute for Policy Studies. During her PhD, she worked at Earth Company as a donor manager and also at Ethiopian Health Insurance Agency of Ministry of Health as a consultant.

Basant Pant is a forestry graduate. He has more than a decade of professional experience working in the issues related with forest and people. Pant is working in the Transboundary Landscapes Programme in the International Centre for Integrated Mountain Development (ICIMOD). He has a master's degree in watershed management from the Institute of Forestry, Pokhara, Nepal, and completed a postgraduate course on 'Protecting Mountain Biodiversity' in 2010, University of Turin, Torino, Italy. Prior to joining ICIMOD, Pant worked as a program officer in Green Governance Nepal, as a rangeland resource person for Northern Mountain Conservation Projects, WWF, Nepal, and as a team leader in two small grant projects funded by the Rufford Maurice Laing Foundation, United Kingdom. He has conducted several academic researches and produced relevant knowledge products in the field of forest and livelihoods. Pant contributes in the book based on primary research conducted in the area. Pant is also a local to the area and has in-depth knowledge on displacement issues and their effects on rural livelihoods. Pant's contribution to the book does not reflect the views of his organization.

Rasyad A. Parinduri is Associate Professor of Business Economics. He does research on the intersection of industrial organisation and development economics—on how policies and market changes affect development, competition, labour outcomes, and trade. His articles have appeared in many journals including the *Economics of Transition*, the *Economics of Education Review*, *The Journal of Development Studies*, and the *World Development*.

Saumik Paul joined ADBI in December 2017. Prior to this, he worked at the Hitotsubashi University, University of Nottingham (Malaysia campus), Osaka University, and the World Bank. He is interested in policy-relevant research on topics related to structural transformation, productivity growth, and land reform. His projects examine land disputes and the process of industrialization in India, Indonesia, and Vietnam, and also the role of structural transformation in regional growth and convergence in Japan and other Asian countries.

Vengadeshvaran Sarma is an assistant professor at the Nottingham University Business School, Malaysia. Sarma's research interests are primarily in the fields of development economics and migration studies.

Naoyuki Yoshino is the dean and CEO of the Asian Development Bank Institute (ADBI); professor emeritus of Keio University, in Tokyo, Japan; and senior adviser at the Japan Financial Services Agency's (FSA) Financial Research Center (FSA Institute). He obtained his PhD from Johns Hopkins University, where his thesis supervisor was Sir Alan Walters (who was Margaret Thatcher's economic adviser). He was a visiting scholar at the Massachusetts Institute of Technology (United States) and has been a visiting professor at various universities including the University of New South Wales (Australia), *Fondation Nationale des Sciences Politiques* (France), University of Gothenburg (Sweden), and Martin Luther University of Halle-Wittenberg (Germany). He was an assistant professor at the State University of New York at Buffalo and an economics professor at Keio University from 1991 to 2014. He was appointed to the Financial Planning Standards Board in 2007 and also served as the chairperson of the Japanese Ministry of Finance's Council on Foreign Exchange and its Fiscal System Council (Fiscal Investment and Loan Program Section). He was also a board member of the Deposit Insurance Corporation of Japan, chairperson of the Meeting of Japanese Government Bond Investors

(Ministry of Finance), and president of the Financial System Council of the Government of Japan. He was conferred honorary doctorates by the University of Gothenburg (Sweden) in 2004 and by Martin Luther University of Halle-Wittenberg (Germany) in 2013; he also received the Fukuzawa Award from Keio University for his contribution to academic research in 2013. He teaches "Financial System of Japan" at the University on Air through the nation-wide television network in Japan every week. He is the president of Financial Education Council organized by the Central Bank of Japan, Financial Services Agency (FSA), Ministry of Education, Cabinet Ministry and private financial institutions.

List of Figures

List of Tables

CHAPTER 1

Introduction

Naoyuki Yoshino and Saumik Paul

"Everything comes from land and ultimately goes back to it" explains an old Indian saying. All sense of value in an enterprise is rooted in the value of land. From business to warfare, the characteristics of land have the power to determine the ultimate success of the endeavour. From maintaining biodiversity to ensuring prosperity for all, the sound management of land resources is vitally important to the achievement of the Sustainable Development Goals (SDGs). The Asian Development Bank notes that more than 400 million Asians still lack electricity; roughly 300 million have no access to safe drinking water, and 1.5 billion lack basic sanitation. In many countries, power outages constrain economic growth. Development of infrastructure is a prerequisite for long-term sustainable economic growth (Yoshino et al. 2018; Banerjee and Duflo 2003; Paul and Sarma 2017). However, more than 8.8% of the GDP is needed to finance infrastructure-related projects in South Asia which are expected to have a spillover on the achievability of the SDGs.

N. Yoshino • S. Paul (✉)
Asian Development Bank Institute (ADBI), Tokyo, Japan
e-mail: nyoshino@adbi.org; spaul@adbi.org

N. Yoshino, S. Paul (eds.), *Land Acquisition in Asia*,
https://doi.org/10.1007/978-981-13-6455-6_1

Infrastructure investment needs in Asia and the Pacific (2016–2030)

	Baseline total	*% of GDP*	*Climate adjusted*	*% of GDP*
Central Asia	33	6.8	38	7.8
East Asia	919	4.5	1,071	5.2
South Asia	365	7.6	423	8.8
Southeast Asia	184	5.0	210	5.7
The Pacific	2.8	8.2	3.1	9.1
Asia and the Pacific	*1,504*	*5.1*	*1,745*	*5.9*

Source: Meeting Asia's infrastructure needs, ADB (2017)

Note: Price in billion dollars for 2015 (annual average)

This book begins by questioning the problems faced in the process of gaining access to land for development purposes. One that dominates this narrative the most is that of Land grab without proper compensation. An issue, which is more prevalent in developing as compared to developed countries, is the imperative to study the effect of land grabbing if one is even to attempt to gain insights into solving the land crisis. Throughout this book, we have focussed on doing just that. We begin with an interdisciplinary approach by combining anthropology and economics. This is done to capture not only the outcome but also the process to help reduce the issue further. This approach not merely allows us to understand the direct livelihood impacts better but also provides insights at the complex ways that displaced communities' capacity for coping with displacement intersects with local historical context over different time periods.

We have also included longitudinal studies by analysing short term versus long term to gauge the adverse impact displacement has on people. The use of longitudinal data in social sciences, particularly in economics, is rare. We have put together a novel panel data set based on our field surveys in Nepal and Indonesia. The case studies concluded that the adverse short-term impact on the displacees normalized over the long term, and the long-term effect could be positive if exact state mechanisms, adequate compensation, and inherent adaptability amongst the displacees were in place. The Asian context is diverse not just regarding various economic structures, but also in the many social hierarchies, which at most are rigid. To further understand these hierarchies and the binding nature they have on social relations, we also studied caste and ethnicity and the impact displacement has on them. Complementing it with a study on the cause or willingness of landowners to sell their land has aided us further in pursuing our multidisciplinary approach.

Having focussed on the above exploratory studies in the first half of the book, we concluded that a robust institutional overhaul is needed to facilitate a smooth and peaceful land transfer. We thus propose the Land Trust Method as a policy framework for more inclusive and secure development. The use of third-party guarantors has a benefit of introducing a check and balance system which we found missing in some of our regional experiences. The Japanese Land Trust Method is explored for feasibility and accessibility in India by employing a legal framework and an inter-temporal analysis. Our findings have encouraged us to pursue this method further by adapting the technique to other Asian countries.

The book is divided into three parts. Part I is dedicated to the conceptual foundations of land grab induced displacement. The second chapter provides an overview of long-term welfare effects of displacement using an interdisciplinary approach. The third chapter studies the short-term versus long-term effects of forced displacement.

Part II focuses on the socio-economic issues and its effects post-displacement, showcasing multiple regional accounts from Asia. The fourth chapter shows the long-term effects of conservation-led displacement in Nepal. The fifth chapter studies the attitudes towards land acquisition in Indonesia whereas the sixth chapter looks at the political economy of caste, displacement, and welfare in Nepal. The seventh and eighth chapters throw light on India, particularly the special economic zones and following livelihood changes as well as specific focus on attitudes towards displacement in West Bengal, respectively.

Finally, Part III proposes the Land Trust Method as a feasible solution and studies its adaptability and accessibility using India as a case study. The ninth chapter introduces the Land Trust Method of Japan and how it may assist in smooth land transfer in other Asian regions. The 10th chapter discusses the Method's accessibility by employing an inter-temporal analysis and spill-over effect whereas the 11th chapter concludes by studying the adaptability of Land Trust Method in India through the existing land pooling structure and its applicability in diverse and non-homogeneous regions.

Chapters' Overview

In Chap. 2, Lam and Paul explore what each discipline of anthropology and economics offers and attempt to identify whether their broken communication can be resolved in particular to the field of displacement studies. Based on the first-hand research regarding the long-term impacts of displacement on local communities, they find that the combination of anthropology and economics offers the most holistic perspective to assess the multi-dimensional effects of displacement because it can capture both the 'process' and the 'outcome' of displacement. The approach not only allows us to understand the direct livelihood impacts better but also provides insights into the complex ways that displaced communities' capacity for coping with displacement intersects with local historical context over different time periods. They use this interdisciplinary model to study long-term livelihood and welfare consequences of forced displacement in Asia in subsequent chapters.

In Chap. 3, Khatua and Sarma examine the short-term versus long-term effects of displacement. In particular, the study explored development-induced displacement and summarized evidence of its short-term and long-term effects from around the developing world. Evidence in the literature points out to adverse short-term effects among displacees that normalize over the long term. In the long term, however, adaptability among displaced and state mechanisms may help displacees normalize and settle-down especially if adequate compensation policies are sanctioned.

In Chap. 4, Lam, Pant, and Sarma focus on the long-term livelihood effects of displacement using a novel panel data set comprising of households affected and unaffected by the Shuklaphanta National Park Extension Programme in Nepal. This study disentangles the impact of displacement along ethnic lines. The main results illustrate that displaced households diversify their livelihoods and work more to compensate for the loss of traditional livelihood. The evidence indicates that almost all ethnic groups were adversely affected: emotionally, they felt a longing for their original lands and felt a lack of communal spirit. The results also show that it is essential to understand the effect of displacement beyond material well-being and across ethnic groups.

In Chap. 5, Yoshino, Parinduri, and Oishi evaluate the attitudes towards land acquisition in Indonesia. They investigate the relationship between landowners' willingness to lease land for infrastructure projects and the characteristics of the agricultural land, household head, and village. The investigation of the determinants of leasing provides insights for land planning and the negotiation process. They found landlords are less willing to rent out their land if they self-cultivate the plots, if plots have been obtained through inheritance, if plots have been owned for a longer period of time, if the land area is small (particularly in the case of non-wetlands and plots located in their villages), and if the ask price demanded is higher than current prices. For the better facilitation of land acquisition for infrastructure projects, it would be advisable to take the above factors into account, as it would encourage landowners to accept the proposed compensation.

In Chap. 6, Lam, Pant, and Sarma explore the deep-rooted ethnic/caste divisions in Nepal and the nexus of displacement and ethnic divisions on welfare heterogeneities since displacement. The main results indicate that there is no difference in the effect of displacement across ethnic lines. However, displaced households fared worse than non-displaced households across all ethnic groups regarding economic and emotional welfare.

In Chap. 7, Paul and Sarma examine livelihood changes of households affected by special economic zones led displacement using a novel data set comprised of households displaced due to the setup of the Falta Special Economic Zone (FSEZ) in India and households indirectly affected by FSEZ. The results indicate lower labour market participation among members of affected households. The results also suggest lower returns in education for displaced households working within the FSEZ, particularly women. However, once pre-FSEZ characteristics are controlled to address endogeneity, the outcomes for displaced households become similar to that of non-displaced households.

In Chap. 8, Paul and Sarma, continuing from the forced displacement experiences of those in the vicinity of the FSEZ in West Bengal, India, herein identify factors that contribute to the willingness to (re-)move. Their study used data generated from a unique household survey administered around the FSEZ region, where people who were already forced to move and those who were not were consulted about their willingness to relocate. They found support for residential land transfer to be stronger among those with property rights, but stronger social capital and current wealth dampens it. Overall, they posit that the propensity to move and expectations for compensation are lower for those with prior displacement experiences.

In Chap. 9, Yoshino, Paul, Sarma, and Lakhia consider the recent political upheaval transgressing from the industrialization drive to violent land-grab in many parts of Asia (India, Indonesia, Nepal, and the Philippines in particular). They posit the need for a sustainable policy—a framework that results in a decisive sum game, benefiting the landowners without hurting the growth prospects. Combining the tools from the fields of law and economics, they propose land trust or land lease for the development of infrastructure investment and industrialization purposes. They argue that this is one of the best ways to increase the rate of return to invite private investors into infrastructure investment. Through evidence from the success stories of land trust initiatives in many Asian countries, they showcase the relevance of this alternate method.

In Chap. 10, Bharule evaluates economic development and its effect on land, focussing on the sequence of decisions which create externalities over the period that either helps in attaining the envisioned development or completely changes the form of spatial development. The chapter explains the process of adopting specialized transport infrastructure like the Mumbai Pune Expressway in case of a developing country like India and the decision-making which went in the process of selecting such infrastructure projects known as road development plans.

In Chap. 11, Lakhia explores the applicability and feasibility of the Land Trust Method in India by studying the region's tumultuous relation with land laws. The chapter considers the existing land pooling structure in India as a probable precursor to the Land Trust Method and identifies likely stakeholders and their roles to facilitate the Land Trust Method within the region.

This book is aimed not just for policy and academic experts within Asia and the Pacific in the diverse fields of economics, anthropology, sociology, international and Asian development studies, displacement studies, and applied economics, amongst others; but also for the emerging leaders of tomorrow—the graduate and undergraduate students from the fields mentioned and more. We hope this book will engage with its multidimensional approach and intrigue those interested in understanding the question and consequences of land displacement within the Asian context.

References

ADB. (2017). *Meeting Asia's infrastructure needs.* Manila.

Banerjee, A., & Duflo, E. (2003). Inequality and growth: What can the data say? *Journal of Economic Growth, 8*(3), 267–300.

Paul, S., & Sarma, V. (2017). Displacement, compensation and channels of endurance: An industrialization-led development saga. *Oxford Development Studies, 45*(3), 240–259.

Yoshino, N., Paul, S., Sarma, V., & Lakhia, S. (2018). Land acquisition and infrastructure development through land trust laws: A policy framework for Asia. *ADBI Working Paper 854.* Tokyo: Asian Development Bank Institute. Retrieved from https://www.adb.org/publications/land-acquisition-and-infrastructure-development-throughland-trust-laws.

PART I

Conceptual Foundations: Displacement and Welfare

CHAPTER 2

Interdisciplinary Approach to Long-Term Welfare Effects of Displacement

Lai Ming Lam and Saumik Paul

Introduction

Economics and anthropology have shared a history of little interdisciplinary interaction. This lack of cordiality has accounted for limited cross-disciplinary publications between them. It is surprising that although these disciplines have forged harmonious relationships with other sister disciplines such as psychology and law, communication between economists and anthropologists remains strained (Cosgel 2005). It is therefore the subject of this chapter to explore what each individual discipline offers and attempt to identify whether their broken communication can be resolved in particular to the field of studies on the long-term effects of forced displacement. The reason we chose the topic on long-term effects of displacement is because we have been conducting an empirical inquiry of the effect of displacement since 2004; our first-hand experiences and observations can therefore provide invaluable contribution to current displace-

L. M. Lam
Osaka University, Suita, Japan

S. Paul (✉)
Asian Development Bank Institute (ADBI), Tokyo, Japan
e-mail: spaul@adbi.org

N. Yoshino, S. Paul (eds.), *Land Acquisition in Asia*,
https://doi.org/10.1007/978-981-13-6455-6_2

ment studies and development studies that particularly looks at the long-term effects of displacement in both theoretical and policy levels.

Although economics and anthropology have been two major disciplines for studying displacement, their coordination was surprisingly rare. This problem was addressed by Michael Cernea, a well-known displacement study scholar who established the 'Impoverishment Risks and Reconstruction' (IRR) model for assessing the impacts of displacement. Cernea (2003) argues that *anthronomics* is not just required to add vigour but is tantamount to the field of displacement and resettlement. He points out several vital issues in *The Economics of Involuntary Resettlement: Questions and Challenges* and *Risks and Reconstruction: Experiences of Resettlers and Refugees*. Primarily, he notes that the economics of displacement has been largely ignored and is thus an immature subject. He further states that anthropology has progressed far ahead of economics in this respect. Thus, meaningful insights can be obtained if displacement and resettlement issues were analysed with an anthro-economic eye.

Moreover, the economic methodology employed to analyse costs, compensation and other displacement issues is based on cost-benefits analysis which neglects and underestimates the full impact of this phenomenon. Indeed, the "compensation principle" has been widely challenged as it fails to truly compensate those displaced (Cernea 2003, p. 10; Kanbur 2003; Schmidt-Soltau 2002; Guha 2001; Nayak 2000). This is because it does not adequately reflect the social and economic losses incurred by displaced communities. This includes not just economic risks but also a breakdown of social networks and relationships (Sapkota 2001; Cernea 1997). Hence, economics alone leads to weak policy recommendations and remedies. An interdisciplinary approach, however, could yield sounder policy and methodological recommendations as well as a truer assessment of the impact of displacement (Cernea 2003). Indeed, it could lead to "double sustainability" to "protect both the biodiversity and people's livelihoods at the same time" (Cernea and Schmidt-Soltau 2006, p. 1808).

The chapter is organized into three parts. In the first part, a review of economics and anthropology is provided. Through the review, we take a closer look at the similarities and differences between these two disciplines. Afterward, in the second part, we question whether methodological differences make them impossible to work together. Indeed, on the contrary, we argue the differences of these two disciplines offer one of the most holistic perspectives to study the complex effects of displacement. Our analysis demonstrates that economics (outcome-focused research) and anthropology (process-focused research) allow us have a better understanding of displacement. In the last part, we use our Nepal displacement study to illustrate the above observations and arguments.

Economics and Anthropology: A History of Little Interaction

Several studies reveal that economics and anthropology have not been the greatest allies. Pieters and Baumgartner (2002) employ citation analysis to examine whether these disciplines communicate with each other. Their findings suggest limited conversations between the aforementioned disciplines. Rigney and Barnes (1980) who use a random sample of articles from one journal for each discipline find results consistent with this. Economics and anthropology, both of which belong to the umbrella of social sciences, share certain distinct commonalities. The most fundamental point where these disciplines meet is their concern for human behaviour and institutions (Cosgel 2005). Despite this, considerable differences between them have precluded cross-fertilization. Extant literature has attributed this to methodological differences. However, Cosgel (2005) contends that behavioural assumptions and modes of enquiry account for this. Thus, while economists assume that rationality is a basic tenet of human behaviour, anthropologists adopt a more holistic approach (Cosgel 2005). It ventures beyond treating man as a utility-maximizing being, taking into consideration a range of motives including social, cultural and moral determinants of behaviour (Cosgel 2005). Furthermore, with respect to modes of enquiry, economists use models, theories and secondary sources to collect quantitative data quite unlike anthropologists who study actual people and events using field research to gather qualitative data (Wilk 1996). The disciplines have been further differentiated based on their purpose—while the former deals with the problem of interest, the latter preoccupies itself with the problem of knowledge (Cosgel 2005).

Much of the differences between anthropology and economics in their approach results from the historical paths they have taken (Buckley and Chapman 1996). Social anthropology has traversed the path of postmodernism wherein it has gone through the positivist phase and abandoned it (Buckley and Chapman 1996). On the other hand, economics has been characterized by modernism, remaining steadfast to its modernistic agenda (Buckley and Chapman 1996). These differing evolutionary paths have resulted in considerable divergences between these disciplines. However, these dissonances do not necessarily prevent economics and anthropologists wholly from talking with one another as cross-disciplinary work abounds among other fields which share similar divergences. Indeed, we argue that significant synergies between economics and anthropology appear promising. Moreover, the flexibility of academic boundaries and the constant state of flux between such boundaries suggest that such differences are not completely irreconcilable and certainly warrant further study.

Benefits from Cross-disciplinary Interaction

There is a growing body of literature that points towards the great rewards that cross-disciplinary study can yield. Economics and anthropology are no exceptions to this. Economics has traditionally studied the more quantifiable dimensions of society with the aid of theories and empirical analysis (Grossbard 1978). On the other hand, anthropologists have flirted with the extensive margin of social science studying entire cultures and communities to gain in-depth insights (Grossbard 1978). The very differences between these disciplines suggest that there is scope for them to work hand in hand to achieve a more informed picture of their subject of study. Indeed, by extending the intensive and intensifying the extensive, the robustness of empirical studies and theories can be combined with wide cultural findings (Grossbard 1978). According to Grossbard (1978), three key benefits can be exploited from such interdisciplinary work: (1) tackling the most pertinent questions more effectively by making use of their respective skills; (2) collecting improved data; and (3) providing new meaning and significance to previous ethnographic findings. Yet, this has seen little application perhaps to "discourage turbulence at the boundaries of a subject out of fear of losing autonomy" (Douglas 1973, p. 781).

Hackenberg (1999) promotes the rather underutilized interdisciplinary hybrid between economics and anthropology. His argument rests largely on Stern's (1960) findings: "Hybridization between different species… often results in increased size, productiveness, and resistance to…unfavorable conditions of the environment." By extension then, if these disciplines cross-breed, significant hybrid vigour could be achieved (Hackenberg 1999). It is thought to occupy particular importance in the field of development planning wherein the resettlement of displaced populations is of major concern (Hackenberg 1999). In this context, economists can form an alliance with anthropologists to ensure that such displaced populations are economically as sound as they were in their previous environs (Cernea 1999). Furthermore, observations by anthropologists are also deemed necessary to come up with more effective solutions. This is one instance of the 'hybrid vigour' that can be achieved through cross-disciplinary interaction. Epstein (1975) purports that the domain of economic development is a perfect ground to marry economists and anthropologists. Indeed, deductive reasoning rooted in the economic man model fails to completely resolve issues such as unequal income distribution (Epstein 1975). This has heightened practitioners' consciousness of the lacunae in the discipline. In what is called a "marriage of convenience", such issues can be addressed and greater explanatory power achieved (Epstein 1975, p. 33). The blend of anthropologists' micro-approach analysing microcosms with the economists' macro-approach acknowledges the importance of both qualitative and quantitative inquiry.

The above arguments are reinforced by Seers (1969, p. 5) who states that "The fulfillment of human potential requires much that cannot be specified in purely economic terms." The following example exemplifies this. Epstein (1968) examined the kinship system of the people of Repitok, the system of inheritance as well as the sale of cocoa. According to the traditional patterns of inheritance, a man's sister's son had a primary claim on the money earned from cocoa rather than his own son (Epstein 1975). Thus, to circumvent this situation, they sold the cocoa seeds to independent Chinese traders so that it did not go on record (Epstein 1975). Thus, their sons could benefit from the money earned. It was only through the analysis of the conflicting pulls from the system of inheritance on one end and a man's loyalty to his son on the other that these insights could be gained. Thus, such interdisciplinary work is of increasing importance especially with respect to development problems. It is therefore fitting that Lipton (1968, p. 14) says "It has become commonplace that macro-plans without micro-knowledge just do not get implemented and are therefore bad plans."

Ray (2006) neatly sums up the benefits that can be reaped by way of cooperative interdisciplinary conversations. Primarily, analytical models are deemed to have better explanatory power if viewed with an anthropological eye. This is because "basic trouble is that nature is so complex that many quite different theories can go some way to explaining the results" (Crick 1988, p. 141). Thus, a particular outcome can be a result of several processes. This is where an anthropologist can add value by examining the processes and structure that led to such an outcome and thereby suggesting alternative explanations for the same outcomes (Ray 2006). One method employed by anthropologists to carry out "empirical investigations of processes themselves" is through within-case causal process analysis wherein each case is studied individually and used to evaluate competing hypotheses (Lipton 1992; Collier et al. 2004).

Moreover, Herskovits (1941) points to lacunae inherent in both disciplines that can be considerably overcome if they joined hands. Economics, based on models and carefully defined assumptions, are considered to be rooted in "logical unreality" (Herskovits 1941, p. 272). It is claimed that findings from such models should be verified by actual facts. This is predominantly the domain of anthropologists who study the actual happenings and rely less on modelling. At the same time, it is suggested that anthropologists adopt a more quantitative approach to thought and practice alike (Herskovits 1941). Furthermore, economic models rest on what is explicitly said (Ray 2006). However, anthropologists believe that what is not being said and asked is as critical as what is being said and asked (Li 2002). This could have political implications and thus more value can be gained if such factors are considered when interpreting a model's outcome. In essence, modelling leads to real gains and losses and the latter can often outweigh the former especially when modelling complex systems (Krugman 1995). It is thus of immense importance that when policies are designed, they incorporate explicit findings from models as well as their silences (Ray 2006). In this context, working with anthropologists could potentially refine the outcomes of such models.

Moreover, it is suggested that economic outcomes aid anthropological enquiry. Economic models often yield counterintuitive or unexpected results (Ray 2006). Such results present an opportunity to anthropologists to investigate new or changing social processes (Ray 2006). For example, Bardhan (2000) finds that there is a U-shaped relationship between intra-village inequality and cooperation over water contrary to expectations. Such instances are an opening for anthropologists to explore the processes and dynamics behind these surprising results. As Appadurai (2004, p. 63) puts it, economic models can often act as an "invitation to anthropology to widen its conceptions of how human beings engage their own futures".

The last two decades have seen a confluence in the interests of anthropologists and economists due to an increase in consciousness of the gains that could be realized (Greenfield 1982). In Gudeman's (2001) book, 'The Anthropology of Economy', he states that recent changes in the global landscape have forced students of economics to rethink their analytical approaches. Indeed, he demonstrates this with the aid of 50 cross-cultural examples thereby presenting an anthropological approach to economy wherein communal processes play a central role in markets (Gudeman 2001). In essence, the book proposes a new medium to tackle contemporary issues more effectively. Lipton (1992) concludes that for a true conversation—*sacra conversazione*—between economists and anthropologists to occur, they must necessarily come to terms with the difference between process-oriented research in anthropology and outcome-oriented research in economics. Processes are the steps whereby specific outcomes are reached whereas outcomes are the research findings obtained from models (Ray 2006). Thus, while economists test whether the modelled process is consistent with the measured outcomes, anthropologists explore the structure of these relationships (Lipton 1992). This is seen to be the key point of departure between these disciplines. For these disciplines to cooperate then, an implicit recognition of the need for both process- and outcome-oriented research is required.

Hence, a better understanding of social phenomena calls for a combination of outcomes from economic models and process analyses of anthropology (Fig. 2.1). This requires economists to examine the outcomes of their models with more caution and anthropologists to explore the possi-

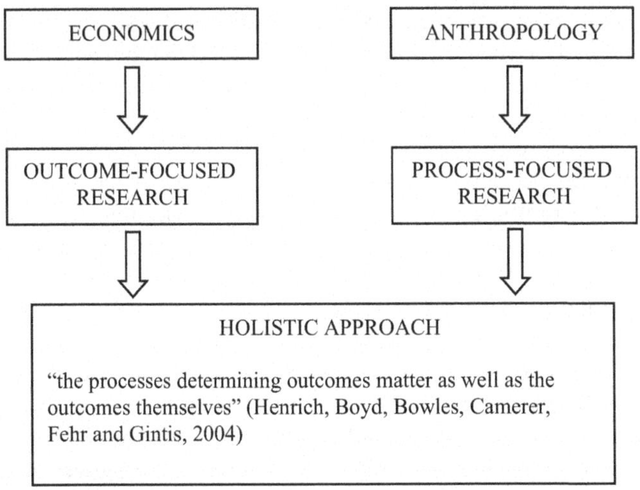

Fig. 2.1 Anthropology and economics

ble usefulness of models (Ray 2006). Perhaps it is fitting to end with Herskovits's (1941) take on the matter—neither can anthropologists ignore the methodological and conceptual tools employed by economists nor can economists ignore the telling contributions anthropology can make to matters such as prestige, gift exchange and specialization of labour. In the ultimate analysis, significant synergies can be gained and thus a more refined body of research attained.

Anthronomics Approach to Study the Long-Term Effects of Displacement in Nepal

Many studies show that since the 1950s, Nepal has experienced rapid transformation in landownership from the indigenous economically marginalized groups who had the weakest political standing to the more powerful immigrant groups (Caplan 1970; Guneratne 2002). It also closely corresponds to the local socio-economic context, particularly after the large number of immigrants led to disruptive social conflicts between indigenous and migrant groups. The backdrop of this socio-economic upheaval involved a displaced indigenous group, the Rana Tharus (hereafter referred to as Ranas), in the western-most districts of Kanchanpur. They experienced large-scale displacement due to the expansion of the Shuklaphanta Wildlife Reserve (hereafter referred to as the Park) in 2001. The Nepalese government carried out a land-based resettlement scheme. It was designed on the principle that all displaced families should be given cultivable land, which they lost previously due to the extension of the wildlife reserve (Bhattarai 2001, p. 270).

Our fieldwork was motivated by the lack of evidence concerning the socio-economic impacts of conservation on marginalized social groups. We adopted a multiple research methodology including household surveys, focus group discussions, participant observations and in-depth participant interviews. While the household survey was designed to capture a broader picture of the socio-economic conditions of the Rana society, the conventional anthropological techniques of participant observations and in-depth participant interviews were conducted to analyse more closely the daily livelihood practices of Ranas and the transformations in the Rana society during the relocation and in the new settlement (Lam 2003). Discussion group participants included local leaders, ex-government officials and local people (both Ranas and hill migrants). Based on repeated consultations with the Park authorities and some local NGOs, the indigenous Ranas from the Rauteli Bichawa Village were considered to be the most appropriate subject of our study.

The Rauteli Bichawa Ranas had to endure many new challenges and the ways in which they cope with those are central to our analysis. This provides us with the opportunity to probe the influence of forced displacement and transformation in landownership on the livelihood of indigenous Rana communities. Three field trips were conducted over a period of 18 months between 2004 and 2006. The sample was restricted to a group of 72 households due to financial constraints and adverse socio-political conditions. The comparison group, comprising 30 Rana households, was selected from the two hamlets of Rauteli Bichawa village, Iymilia and Jhimila, located near the periphery of the Park. The resettled group selected for our study included 42 displaced Rana households from the two hamlets, Rampur and Beldandi of the Dhokka Block (Table 2.1).

Both quantitative and qualitative approaches were used to evaluate the long-term welfare effects of forced displacement on Rana households. We first discuss the quantitative results. To determine the overall welfare impact we follow a simple regression-based approach. The mean comparison approach is appropriate in our case because the resettled group can be identified based on observables. The dependent variable in Table 2.2 is the welfare indicator measured as food security in the future measured in months. We decided not to use productivity (measured as crop yield) as the dependent variable because it is an estimation based on focus group discussions. Despite the fact that data on food security suffer from self-reporting bias, data directly reflect the availability of food after resettlement and their understanding of what food sufficiency means at the local

Table 2.1 The number of Rana households in the four study settlements

	Rauteli Bichawa Village		*Dhokka Block*	
	Iymilia	*Jhimila*	*Rampur*[a]	*Beldandi*
Total households	100	165	506	460
Rana households	90	20	126	19
Surveyed households	15	15	25	17

Source: Household Survey 2005

[a]The Rampur estimate was based on information provided by the ex-chairperson of Beldandi and Rampur Buffer Zone User Group Committee, Bhim Thapa

Table 2.2 Regression Outcomes on Food Security

	Dependent variable: Food security (months)	
	Base plus household control	*Restricted sample*
	(2)	*(3)*
Resettled (yes = 1)	−1.71*	−2.00*
	(0.73)	(0.83)
Log age	0	−0.43
	(0.56)	(1.47)
Gender (female = 1)	1.55*	1.28
	(0.74)	(0.88)
Household members	−0.08	−0.12
	(0.05)	(0.07)
Land holding (in katta)	0.03**	0.03***
	(0.01)	(0.01)
Livestock (numbers)	−0.01	0.32
	(0.11)	(0.22)
Constant	8.89***	9.81
	(2.28)	(5.43)
Observations	67	53
R-square	0.2	0.24

Note: Restricted sample implies only those households who resettled in 2001

Robust standard errors are given within parenthesis. Coefficients with * mean significant at 10%, ** mean significant at 5% and *** mean significant at 1%

level. In addition, the Rana respondents did not include incomes from non-agricultural activities while answering the food security question. They only considered the family size and the amount of crop yield from their own cultivation. Together with this, the food security indicates the productivity and family size as well.

In Table 2.2, we provide regression outcomes on food security. Displaced households are found to have lower food security irrespective of the way it is measured. The outcome is robust and statistically significant in most cases. The models with the food security variable measured in terms of the number of days they have enough food given the landholding size show better fit. While smaller households are better off, the households with bigger land on average have food security for a longer period of time. For the purpose of robustness, we ran the same model on a restricted sample comprising only those Rana households who resettled in 2001. Overall, the outcome remains unaffected.

While quantitative evidence suggests that by and large the land-based compensation policy has failed to prevent impoverishment in the Rana society, it does not provide much evidence on the social injustice that Ranas faced over a long time. The Ranas, like many traditional societies, failed to perceive the modern concept of landownership as an exercise in land registration documents. For them, the concept of landownership was more about the actual land use practices. Guneratne (1996, 2002) explains that the concept of obtaining the legal land documents to secure ownership does not exist among many tribal or ethnic communities, particularly those from the lowland Tarai region of Kanchanpur. In focus group discussions, many Rana informants mentioned that they had been cultivating their land for generations so they never feared losing it. This, however, put the indigenous Rana population into a weak position to protect their ancestral land, particularly those with small landholdings.

The story of Jekur Rana provides an example. The Jekur Rana family is one of the displaced families from the Andaiya hamlet of the Rauteli Bichawa Village. He had 100 Kattas of ancestral land, which had been used as the main source of livelihood through subsistence agriculture for more than 100 years. However, his land was not registered officially. According to him, the older generations had no idea about the land registration procedure. Moreover, when government officials came to their village on one occasion, they only talked to the rich and educated people, not them. As a result, only the rich and influential families, including some wealthy Ranas, registered their land with the government. In 2001, the family of Jekur Rana was forced to move out from the extension area of the Park. Since he did not possess any legal registration, Jekur Rana's family received only two Kattas in order to build a shelter in the new resettlement area in accordance with his inhabitant status. Jekur Rana pleaded to the Park authority to reassess his case many times but without success. There were at least ten other Rana families in the Dhokka Block in a similar situation like Jekur Rana.

Focus group discussions with both resettled and non-resettled Ranas found that the Ranas, who had close relations with local elites and owned large plots of land obtained official documents and thus suffered less from the relocation. As our data show, a majority of the displaced Ranas receiving almost equivalent size of their registered land were rich, owning more than 200 Kattas of land inside the Park. Thus, the design of the state policy of the land compensation scheme apparently favoured the rich and it only increased social inequality by impoverishing the poor at a higher rate.

Since the 1950s, the state has played a leading role in the transformation of landownership from the hands of indigenous Ranas to migrants from the hills (Pahaaris). This was administered through a series of land reform policies and state-sponsored resettlement programmes in Kanchanpur, particularly in the Rauteli Bichawa Village and in the Tarai region as a whole. The migrants were mainly higher caste people, including Brahmans and Chhetris. They were mostly literate and had closer ties with the state officials, such as sharing the same language (ability to speak and write in Nepali) and culture. This made access to land resources and assistance from the state easier for them and in turn gave them greater control over land.

As found in the focus group discussions, this was apparently another major reason why some Ranas could not register their land properly or even lost most of their land to the migrants. Many Ranas complained that in many instances the disputes over land between them and the migrants were resolved in favor of the migrants. As in matters regarding the registration and transaction of land, it required good communication skills with the state officials verbally and literally. There were also complaints against the migrants that they took advantage of the illiterate Ranas and confiscated their land by providing them with flawed contracts. For example, one displaced Rana stated that without the consent and authorization of his grandfather, his father signed a land transaction document to a migrant state official. However, when his grandfather contested it, the land was already a property of the state official.

As our qualitative evidence suggests, the impact of Nepal's land compensation policy has resulted in a disproportionate distribution of land where the poor have come out the worst. This has serious consequences for the social deprivation of marginalized groups that have less political clout. Overall, both quantitative and qualitative evidences suggest the necessity of a land compensation framework that must consider overcoming the social divisions and political economy of past land settlement policies. Without thoughtfully considering the political, economic and cultural contexts, land-based compensation schemes will only serve as a mechanism to further accelerate social inequality and social strife among different groups. We discuss these issues in further detail in the subsequent chapters.

References

Appadurai, A. (2004). The capacity to aspire: Culture and the terms of recognition. In V. Rao & M. Walton (Eds.), *Culture and public action* (pp. 59–84). Stanford, CA: Stanford Social Sciences.

Bardhan, P. (2000). Irrigation and cooperation: An empirical analysis of 48 irrigation communities in South India. *Economic Development and Cultural Change, 48*, 847–866.

Bhattarai, A. M. (2001). *Displacement and rehabilitation in Nepal: Law, policy and practice*. New Delhi: Anmol Publications.

Buckley, P. J., & Chapman, M. (1996). Economics and social anthropology—Reconciling differences. *Human Relations, 49*, 1123–1150.

Caplan, L. (1970). *Land and social change in east Nepal: A study of Hindu-tribal relations*. London: Routledge & Kegan Paul.

Cernea, M. M. (1997). The risks and reconstruction model for resettling displaced populations. *World Development, 25*(10), 1569–1587.

Cernea, M. M. (1999). Why economic analysis is essential to resettlement. In M. Cernea (Ed.), *The economics of involuntary resettlement* (pp. 5–49). Washington, DC: World Bank.

Cernea, M. M. (2003). *For a new economics of resettlement: A sociological critique of the compensation principle*. UNESCO, Paris: Blackwell.

Cernea, M. M., & Schmidt-Soltau, S. (2006). Poverty risks and national parks: Policy issues in conservation and resettlement. *World Development, 34*(10), 1808–1830.

Collier, D., Mahoney, J., & Seawright, J. (2004). Claiming too much: Warnings about selection bias. In H. Brady & D. Collier (Eds.), *Rethinking social inquiry: Diverse tools, shared standards* (pp. 85–102). Lanham, MD: Rowman & Littlefield.

Cosgel, M. (2005). Conversations between anthropologists and economists. *Economics Working Papers 200529*. Retrieved from http://digitalcommons.uconn.edu/econ_wpapers/200529

Crick, F. (1988). *What mad pursuit: A personal view of scientific discovery*. New York: Basic.

Douglas, M. (1973). The exclusion of economics. *Times Literary Supplement*, pp. 781–782.

Epstein, T. S. (1968). *Capitalism, primitive and modern—Some aspects of Tolai economic growth*. Canberra: ANU Press.

Epstein, T. S. (1975). The ideal marriage between the economist's macroapproach and the social anthropologist's microapproach to development studies. *Economic Development and Cultural Change, 24*(1), 29–45.

Greenfield, S. M. (1982). Anthropology and institutional economics. *Journal of Economic Issues, 16*(2), 485–487.

Grossbard, A. (1978). Towards a marriage between economics and anthropology and a general theory of marriage. Papers and Proceedings of the Ninetieth Annual Meeting of the American Economic Association. *The American Economic Review, 68*(2), 33–37.

Gudeman, S. (2001). *The anthropology of economy: Community, market, and culture.* Oxford: Wiley-Blackwell.

Guha, A. (2001). *Land acquisition among the cultivators of rural Medinipur, West Bengal: Anthropological appraisal.* PhD dissertation, Vidyasagar University, India.

Guneratne, A. (1996). The Tax Man Cometh: The impact of revenue collection on subsistence strategies in Chitwan Tharu Society. *Studies of Nepali History and Society, 1*, 5–35.

Guneratne, A. (2002). *Many tongues, one people: The making of Tharu identity in Nepal.* New York: Cornell University Press.

Hackenberg, R. A. (1999). Advancing applied anthropology—Victims of globalization: Is economics the instrument needed to provide them a share of the wealth? *Human Organization, 58*(4), 439–442.

Herskovits, M. J. (1941). Economics and anthropology: A rejoinder. *Journal of Political Economy, 49*(2), 269–278.

Kanbur, R. (2003). *Development economics and the compensation principle.* UNESCO, Paris: Blackwell.

Krugman, P. (1995). *Development, geography and economic theory.* Cambridge, MA: MIT Press.

Lam, L. M. (2003). *Change of perceptions of local communities and park-people conflicts in regard to Koshi Tappu Wildlife Reserve, Nepal.* Unpublished master dissertation, The University of Adelaide, Adelaide, Australia.

Li, T. M. (2002). Engaging simplifications: Community-based resource management, market processes and state agendas in upland Southeast Asia. *World Development, 30*, 265–283.

Lipton, M. (1968). Interdisciplinary studies in less developed countries. *Journal of Development Studies, 4*, 15.

Lipton, M. (1992). Economics and anthropology: Grounding models in relationships. *World Development, 20*, 1541–1546.

Nayak, R. (2000). In M. Cernea & C. McDowell (Eds.), *Risks associated with landlessness. Risks and reconstruction.* Washington, DC: The World Bank.

Pieters, R., & Baumgartner, H. (2002). Who talks to whom? Intra and interdisciplinary communication of economics journals. *Journal of Economic Literature, 40*(2), 483–509.

Ray, I. (2006). Outcomes and processes in economics and anthropology. *Economic Development and Cultural Change, 54*(3), 677–694.

Rigney, D., & Barnes, D. (1980). Patterns of interdisciplinary citation in the social sciences. *Social Science Quarterly, 61*, 114–127.

Sapkota, N. (2001). Impoverishment risks and reconstruction of Kali Gandaki Dam, Nepal. *High Plains Applied Anthropologist, 21*(2), 147–156.
Schmidt-Soltau, K. (2002). *The environmental risks of conservation-related displacements in central Africa*. Paper prepared for the International Symposium on Multidimensionality of Displacement Risks in Africa, Kyoto, Japan.
Seers, D. (1969) *The meaning of development* (IDS Communication Series 44) (p. 5). Brighton: Institute of Development Studies, University of Sussex.
Stern, C. (1960). *Principles of human genetics*. San Francisco: Freeman.
Wilk, R. R. (1996). *Economies and cultures: Foundations of economic anthropology*. Boulder, CO: Westview Press.

CHAPTER 3

Short-Term Versus Long-Term Effects of Forced Displacement

Sayak Khatua and Vengadeshvaran Sarma

Introduction

Many developing economies have been following a policy of large-scale public sector investment for providing infrastructure and state investments for development—these include roads, dams, special economic zones and so on. These projects have caused major changes in the use of land and other natural resources and thereby increased the productivity of land—positively contributing to the gross domestic product (GDP) of many economies. At the same time, they have caused irreversible changes in the lives of thousands of people who had inhabited or depended on those 'green' land and other natural resources for their survival. The acquisition of land for development involuntarily displaces the local inhabitants, mostly the poor and the deprived sections of the population. Given the scale of growth in many developing countries especially in Asia and sub-Saharan Africa, the magnitude of involuntary displacement is likely to increase over the years (see International Migration Report 2017). As

S. Khatua
International Initiative for Impact Evaluation (3ie), New Delhi, India

V. Sarma (✉)
University of Nottingham, Semenyih, Malaysia

N. Yoshino, S. Paul (eds.), *Land Acquisition in Asia*,
https://doi.org/10.1007/978-981-13-6455-6_3

more people are forced to leave their homes and traditional livelihoods for projects that have benefits for the greater economy, this chapter focuses on the theory and evidence of displacees' experiences.

Theorising Forced Displacement

Anthropologists identify various push and pull factors due to which people move from one place to the other. The push factors comprise economic, political, environmental and social reasons. Push and pull factors are often unique to specific regions; a few general examples of the factors are listed in Table 3.1.

These factors consider both the voluntary and involuntary aspects of migration. Forced displacement is the result of push factors that generally force individuals out of their indigenous localities and make them move to a new place. An individual, when pushed out of his or her own comfortable habitat, is faced with many uncertainties. Involuntary resettlers are generally more anxious and face higher levels of insecurity compared to other migrants (Cernea and Guggenheim 1993).

Forced or involuntary displacement is a multi-faceted issue that has affected people in the recent past. People are forced to migrate due to conflicts, environmental factors, natural disasters and man-made projects. Forced displacement can be looked at through multiple lenses—sociology, anthropology, economics, health, public policy and so on. Due to this multi-faceted nature, forced displacement is difficult to model theoretically. Over the past 50 years or so, several theoretical models have been developed to explain issues surrounding displacement. Most of these models deal with the idea of resettlement. Resettlement has been majorly addressed by these models as the immediate impact of displacement is on livelihoods and shelter.

Table 3.1 Examples push and pull factors of migration

Push factors	*Pull factors*
Poverty	High wages
Unemployment	Improved standard of living
Low wages	Professional development
High fertility rates	
Education	

In this section, we discuss a few seminal models to understand the evolution of theoretical models that have been put forth over the last few decades.

Chamber's Three-Stage Model

Chambers (1970) identified a three-stage model of evolution of land resettlement policies. He identified the stages as recruitment, transition and development. Empirical studies such as that of Nelson (1973) (and studies cited therein) that evaluated development-induced displacement (DID) issues in Latin America confirm the applicability of Chamber's model. The model developed by Chamber and expanded by Nelson generalised the experience of *voluntary settlers* and thereby managed to conceptualise the institutional dimensions of managed land settlement programmes.

John Rawls' General Conception of the Justice Model

Hare and Rawls (1973) states that all social primary goods such as liberty and opportunity, income and wealth and the bases of self-respect are to be distributed equally unless an unequal distribution of any or all of these goods is to everyone's advantage. When we talk about forcefully displaced populations, Rawls' theory can be used to identify the problems encountered by the displacees, that is, the negative impacts that forced displacement has had on them. In the long term, if these displacees are not resettled with proper economic means there is a greater chance of these populations being impoverished.

In Rawlsian theory, forced displacement can create inequalities not only of income and wealth but also of social goods in terms of liberty and opportunity and the bases of self-respect. Further, based on Rawls' theory of justice, forceful displacement of people is not only impoverishing but also unjust. Economists talk about compensation for different projects, but these compensations are far from being just. In line with Rawls' theory when a person is displaced, in the short term, this individual needs a community that accepts him or her and provides this person with access to land, to resources for production and other subsistence resources that overall facilitate the individual's new life. Communities should provide enough support for the cultural survival of the displacee as well.

Unbalanced Growth

Most of the issues with forced displacement can be largely attributed to the emergence of the unbalanced growth theory being used for rapid economic development. The unbalanced growth strategy in the developing and under-developed region ignores the marginalised population. DID has become a common occurrence in the developing world, where vulnerable populations are forcefully moved to make way for development projects. Forceful displacement comes with a number of issues that are sometimes ignored by governments or institutions, whose main goal is large-scale development. Literature on this topic points out that most of these large-scale development projects provide inadequate compensation and poor resettlement packages. Furthermore, the loss of assets and compensation often paid in cash create problems as assets are often difficult to rebuild and cash runs out easily. This often makes displacees worse off. The vulnerable population with lower levels of education and income are the most affected as they lack knowledge and resources to rebuild a new life.

The Scudder and Colson Model

Scudder and Colson developed a four-stage model of resettlement based on Chambers' and Rawls' theories. The four stages are recruitment, transition, development and incorporation/handing over. The model looks at the behaviour of resettlers, focusing on the stress of the settlers and their specific behavioural reactions in each of the aforementioned stages.

The model was initially set up to analyse voluntary resettlements but was later adopted to analyse involuntary displacement under certain conditions. When a particular involuntary resettlement is successful and goes through all the four stages of the model, then the resettlement can be analysed under the purview of the Scudder and Colson model. Irrespective of the kind of relocation, voluntary or involuntary, the initial phases are the most stressful in any migrant's life. In the short term, they are faced with unavoidable adversities, adjustment issues and acceptance. These issues resolve with time and in the long term they become more flexible, individualistic and find a place for themselves both culturally and aesthetically in the place of relocation.

The Scudder and Colson model has come under a lot of criticism from different academic spheres. De Wet (2001) talks about how the model ignores the differences in peoples' reactions to involuntary displacement and focuses more on the similarities between voluntary and involuntary relocation. The model incorporates psych which is a step above Chamber's three-stage model but in Cernea's point of view there is more to this issue than just stress. Cernea feels that behavioural patterns of the displaced population need to be followed closely and their situation needs to be looked at with respect to economic, social and cultural impoverishment.

Impoverishment Risk Reconstruction

Cernea (1997) in the Impoverishment Risk Reconstruction (IRR) model shifted the focus from stress to impoverishment of the displacees. The IRR model was developed on the premise that forced displacement has marginalised and impoverished people more than benefited them. The IRR model presents eight potential risk factors associated with displacement. The risks are landlessness, joblessness, homelessness, marginalisation, food insecurity, increased mortality and morbidity, loss of access to common property and social disarticulation.

Cernea based his model on the idea that impoverishment is the main problem that is faced by the forcefully displaced population. The IRR model further explains the cultural and social dimensions apart from the economic aspects of impoverishment. In this model, Cernea lists potential risks such as landlessness, joblessness, homelessness, marginalisation, food insecurity, increased morbidity and mortality, loss of access to common property and services and social disarticulation. The most significant part of the IRR model is about rebuilding the lives of the displacees. Cernea represents a theoretical model of impoverishment due to displacement. The process of rebuilding takes into account all the risks listed above and provides ways to mitigate these risks for displacees. This model has been recognised not only by economists but also by anthropologists and sociologists.

Forced displacement can be theoretically further categorised as a shock to the lives of the people that are displaced. A shock is a phenomenon that brings about a change in the existing livelihood of an individual. Shocks in the form of forced migration damage and destroy important inputs (Fiala 2015) that are used by households for the production of consumables. A one-time shock to capital in the short term will not affect long-term income and consumption, as income is likely to be fungible. This happens as displacement leads to investment in physical and human capital which results in the long-term stability of income and consumption. Given that most of the displacees comprised vulnerable populations with no or limited access to physical and human capital, displacement leads to impoverishment of these populations in the long term. This can also lead to a poverty trap.

A majority of these models were developed out of the need to analyse the effects of voluntary and involuntary displacement on the displacees or host populations. These models mostly discuss the issues related to resettlement and how proper resettlement policies can benefit the displacees. The models often haven't really captured the impact that voluntary or involuntary displacement has on the people involved. There are impacts that are quantitatively not measurable and thus, coming up with a theoretical model is difficult. Cernea's IRR model to an extent identifies the issues and tries to analyse the impact aspect of forced displacement.

Impact of Forced Displacement

Forced displacement affects the displacees in multiple ways. The marginalised and vulnerable groups in the population are the most affected. There are economic, political, social, health and cultural impacts that are permanent in nature. The short-term impacts affect displacees right when they move in the initial stages of coming to terms with the displacement. Long-term impacts develop over time and have an everlasting effect on the lives of the displacees. The effect of forced displacement is mostly apparent on the marginalised and the vulnerable groups. These are people who not only face economic marginalisation but also face psychological and social marginalisation. Bjonness (1983) found that the displacement due to the construction of the Nepal Kulekhani Hydroelectric Project made the displacees worse off socially and economically due to lower productivity of the new lands. Lam and Paul (2013) similarly demonstrate erosion of safety nets and vicious cycle of poverty for displacees of the Shuklapantha Wildlife Reserve Project.

Apart from the marginalised and vulnerable groups, women suffer the most from the impacts of forced displacement. Women spend most of their time at home, building strong social networks which break down when they move. The loss of kinship networks and the dominance of men make their move way worse for women compared to men (see Puechguirbal 2010). Due to development projects, the access to natural resources has become scarce and these women must travel farther to collect fuel and water. This limits their time spent with their kids which in turn affects their family dynamics. In terms of social norms, the marriage and the dowry system are disrupted and there are difficulties in negotiations among families. Also, early marriage practices increase and kids as early as 14–15 years get married. There is an increase in domestic abuse as well. Overall, women face greater adversities than men because of forced displacement (see Benjamin and Khadija 1998).

Internally displaced people (IDP) are widely ignored in the literature on forced migration. Most international standards for resettlement centre around cross-border displacement. The IDP are often ignored. Due to this their exploitation goes unnoticed and human rights violations go under the radar. IDPs don't leave their country of birth, but their struggles are similar to refugees.

The governments justify the displacement of people by encapsulating development as the key to growth, and to achieve growth, dams, roads, factories, infrastructure and so on. need to be built. In the USA, this is captured under eminent domain and similar legal provisions in other parts of the world guarantee that governments justify development-related land acquisition and resultant human displacement as a necessity for the benefit of the nations. Governments view economies as being in pareto-optimal conditions, that you cannot make any stakeholders better without making some others worse off. However, state authorities view development projects as so quintessential that any opposition to these projects often results in harsh measures by the government to suppress dissent—notably in India (see Paul and Sarma 2017). For example, in Kochi during 2006/2007, 38 households were displaced by the Kochi Municipal Corporation to set up a small-scale urban renewable project. The officials from the municipal corporation classified the 38 evicted households as illegal encroachers while the people were living in the same area since 1975 (Walicki and Swain 2016).

Short-Term Effects

Short-term or temporary effects of forced displacement are largely influential in shaping the future of displacees. These are immediate effects that arise when the displacee is forced to move and have an impact in the initial years of the displacees' future. Short-term impacts generally induce psychological stress which is a result of multiple uncertainties. In theory, the impact in the short term is a shock that can manifest as an income shock, cultural shock and so on.

The initial days after displacement are the most crucial. Displacement due to development projects would entail the relocation of the displacees to a location which has been pre-determined by the government or the institution involved. In case of conflict- or natural disaster–led displacement, the displacees find themselves in camps or make-shift shelters. The two situations are completely different when it comes to understanding the impact of displacement on these displacees.

In most cases, people displaced due to development projects, are provided compensation and relocated to a pre-determined location. The government mainly takes onus in relocating displacees in the case of development projects. Although these displacees have a place to stay, they still need much more in terms of rehabilitation. The resettlement of the displaced population is a big challenge as it entails much more than just providing a new home for the displacees. There are certain aspects that take a big hit when people are relocated against their wishes. The biggest impact is on the livelihood of the bread-earner of the family. There are uncertainties of income, savings and people losing their sense of security. The family dynamics change; families with children have the toughest time adjusting to new locations. We often find gender role reversals in the sense that women now become bread-earners doing odd jobs and the men stay at home as it becomes difficult for them to find jobs in the new environment. Fiala (2012) find evidence that displaced individuals face adverse economic conditions. These adverse conditions are mostly the result of dispossession.

Some examples of the short-term impacts of forced displacement on economic outcomes in the literature are discussed here. Kondylis (2010) explores the impact of forced displacement on labour market outcomes in Bosnia and Herzegovina. The data collected were from the people displaced because of the war. The study found that the displacement negatively affected the labour market outcomes for both men and women. Men faced high levels of unemployment with an increase from 16% to 29%. Women predominantly dropped out of the labour force; there was an increase in inactivity of women from 11% to 18%. Fiala (2009) studies the valuation of assets of households and their consumption patterns in Northern Rwanda. He found that the value of assets has no association with displacement, but the households that are not displaced face losses in the valuation of their assets. Further, displacement increases the number of assets that a displaced household had before being forcefully relocated. Ibáñez and Vélez (2008) study consumption patterns and welfare impacts of forced displacement. They found that welfare losses due to displacement decrease in net present value of rural lifetime consumption by 37%.

Health risks have a close association with forced displacement. Most of the literature talks about the impact on health that arises due to displacement resulting from natural disasters, conflicts and wars. Risks that arise due to DID are often ignored. The health of displacees is affected in different ways based on the cause of displacement. Most of the health issues are likely to be intergenerational and spill over into the long term as displacement destroys the coping mechanisms that displacees had. When the displacees relocate, the physical and cultural infrastructure changes and it becomes very difficult for them to feel comfortable both physically and mentally. Stress is one of the most commonly found illnesses that is faced by the displacees.

Data suggest that forced displacement increases the morbidity and mortality rates. Displacees comprising pregnant women, infants and the elderly are affected the most. These people need immediate care and it is very difficult for displacees to access good medical facilities immediately after they have been relocated. For natural disasters or wars and conflicts, one finds international aid agencies providing support both with medical supplies and with infrastructure. But when we look at IDP, the evictor is in itself the caretaker of these people. In other words, the government is responsible for the well-being of the citizen and it is the body responsible for the forced displacement of people.

Long-Term Effects

Economists believe that in equilibrium, the returns to displacement equal the cost of displacement. But sociologists and anthropologists suggest that there are other intangible factors that cannot be quantified in terms of reaching an equilibrium. The risks identified by Cernea (1997) have short-term as well as long-term impacts. The effect of these risks is felt severely in the long term if not addressed in the short term. Without a proper resettlement or reconstruction plan, the impact of displacement on the displacees becomes aggravated with time. In the long term, the displaced population often become further impoverished and face multiple health-related issues.

Sarvimäki et al. (2009) look into the long-term impacts of displacement using displacement in Finland after the Second World War. They looked at outcomes after 50 years of displacement and found that displacement actually increased the long-term income of the displaced. But the authors suggest that these long-term gains were a result of effective reconstruction and resettlement policies. This shows that if given an opportunity, displacees in the process of rebuilding their lives create huge economic impacts on the income of the area they are displaced to. Further, they bring new skills that might become valuable in the long term with demand. Falck et al. (2011) and Bauer et al. (2013) do not find displacees in Germany to be worse off than before their move. Many of the second-generation families still lived in utmost poverty. The studies do find that changing sectors from agricultural to industrial significantly increased the income of the displacees.

Global Evidence

There is a dearth of statistics on the number of people displaced due to development. There is also a dearth of studies that look at cross-country analyses, except for reports from international development organisations (IDOs) such as the World Bank. Unofficial figures suggest that countries like India and China are responsible for a large number of forcefully displaced people. The need for rapid growth has made development necessary at the cost of the vulnerable population. The World Bank Environment Department (WBED) estimates that each year more than ten million people are displaced due to the construction of dams, urban development and transportation and infrastructure projects. In the name of development governments can acquire land legally and displace the indigenous communities with no heed to the effect that it has on the cultural identity of the people.

Asia and Pacific

In this region, there are two giants—India and China. Countries which have divulged immense amount of funds in development in the recent past and currently are fighting to be the world's superpowers. Fuggle and Smith (2000) calculated that in China more than 45 million people have been displaced due to various development projects from 1950 to 2000. In China, the Three Gorges Dam construction displaced over 1.2 million people. As a result of the construction, around 13 cities, 140 towns and 1350 villages were flooded. Further, the project was marred by corruption, human rights violations and difficulties in resettling displacees. The impact of dam construction was fatal to the ecosystem creating a microclimate. China officially acknowledged, in 2011, the Three Gorges Dam as an urgent problem.

India on the other hand has undertaken similar relocations with the Sardar Sarovar Project on the Narmada. This project is responsible for displacing close to 200,000 people. More than half of the people affected by the construction of the dams were *adivasis* (tribal people). The Morse Commission Report (1992) highlighted key factors concerning environment and resettlement issues. The World Bank withdrew funding as there were several concerns about the project.

Africa

Cernea (1997) and De Wet (2001) have spoken about displacement in Africa through the lens of DID and provided in-depth conceptual overview and statistics. Cernea points out that although India and China lead when it comes to number of people displaced due to development projects, the proportion of population displaced in different African countries is much higher for certain projects. For example, the Akosombo Dam in Ghana displaced about 1% of the population whereas the Sardar Sarovar Dam in India displaced about 0.013% (1980) of the population. Further the Akosombo Dam flooded nearly 3.5% of land area of Ghana whereas the Sardar Sarovar Dam only flooded 0.01% of land area.

Colson (1976) did a follow-up study on her 1967 paper on the Kariba resettlement scheme, detailing the Gwembe life. She discusses the pre- and post-scenarios of the displacement and how forced displacement affected the lives of the Gwembe. The Kariba Dam project displaced nearly 57,000 people. Another example would be the displacement due to the construction of the Aswan High Dam. The construction displaced around 100,000 people in Egypt and Sudan, as documented in the work of Fahim (1981). The dam's reservoir affected the nomadic tribes in the Nubian region by destroying their summer resources. Further, the compensation and resettlement schemes failed miserably. Amarteifio (1966) discuss displacement and resettlement of a particular village in Ghana to make way for a new port and harbour.

Latin America and the Caribbean

Displacement in this region is not as severe as in Asia. A few examples of displacements due to development projects are discussed here. The construction of the Tucuri Dam in Brazil displaced about 35,000 people, although the pre-project estimate of the displacement was only 1,750 families (La Rovere and Mendes 2000). In Guatemala, 2,500 Maya Achi Indians were resettled due to the construction of the Chixoy Dam and there was a massacre of 369 people. Howard (1997) talks about the displacement due to the construction of the Peligere Dam and the spread of Green Revolution technologies in Haiti. Mexico's Miguel Aleman Dam displaced nearly 25,000 Maztec Indians. Resettlement policies did not stop the impoverishment of the displaced Maztec Indians (Barabas and Bartolomé 1974).

Europe, the USA and Canada

DID is not common in the USA, Europe and Canada, at least since the Industrial Revolution. There haven't been many studies documenting displacements in these regions. A few notable examples are also small-scale displacements compared to the displacements in Asia and Africa. The James Bay Power Project in Canada displaced the Cree people destroying their unique way of living for generations (Scudder 1996). The Grand Coulee Dam displaced indigenous and non-indigenous families. The project extended for over 40 years displacing over 6000 people. The project not only affected people within the USA but also people in Canada, who did not even receive any compensation (Ortolano 2000). Furthermore, the construction of the Garrison Dam in the 1950s displaced close to 300 indigenous families from protected lands.

An overarching theme that can be highlighted in all these projects is the fact that vulnerable populations are mostly affected. Indigenous tribes who have unique ways of sustainable livelihood get affected the most. The poor become further impoverished due to these development projects. These development projects undertaken by the governments portraying the idea of national progress not only has damaged the environment but also economically affected people in a negative way. These examples show a global picture of the state of the vulnerable populations and how they are exploited.

Conclusion

In this chapter, we focused on the causes and consequences of forced displacement—primarily DID—and summarised some evidence from around the globe. DID tends to be a phenomenon across the developing world, especially in China, India and more recently across sub-Saharan Africa. Governments in developing countries justify displacement as a necessary suffocation of a minority of the people to alleviate growth and employment opportunities through development activities for the entire country. Evidences in the literature highlight that displacees suffer mostly in the short term. In the short term, adverse psychological, income and cultural factors affect individual and family security and tend to make displacees worse off compared to non-displaced households. In the long term, however, adaptability among displacees and state mechanisms may help displacees normalise and settle down especially if adequate compensation policies are sanctioned. State mechanisms that relate to development-led displacement are therefore crucial in ensuring limited harm to displacees.

References

Amarteifio, G. W. (1966). *Tema Manhean: A study of resettlement.* Ghana Universities Press for the University of Science and Technology, Kumasi; Oxford University Press.

Barabas, A., & Bartolomé, M. (1974). Hydraulic development and ethnocide: The Mazatec and Chinantec people of Oaxaca, Mexico. *Critique of Anthropology, 1*(1), 74–102.

Bauer, T. K., Braun, S., & Kvasnicka, M. (2013). The economic integration of forced migrants: Evidence for post-war Germany. *The Economic Journal, 123*(571), 998–1024.

Benjamin, J. A., & Khadija, F. (1998). *The gender dimensions of internal displacement: Concept paper and annotated bibliography.* New York: UNICEF.

Bjonness, I.-M. (1983). External economic dependency and changing human adjustment to marginal environment in the high Himalaya, Nepal. *Mountain Research and Development, 3*(3), 263.

Cernea, M. (1997). The risks and reconstruction model for resettling displaced populations. *World Development, 25*(10), 1569–1587.

Cernea, M. M., & Guggenheim, S. E. (1993). *Anthropological approaches to resettlement: Policy, practice, and theory.* Boulder, CO: Westview Press.

Chambers, R. (Ed.). (1970). *The Volta resettlement experience.* New York: Praeger/ Volta River Authority, Accra and University of Science and Technology, Kumasi.

Colson, E. (1976). *The social consequences of resettlement: The impact of the Kariba resettlement upon the Gwembe Tonga.* Manchester, UK: Published on behalf of the Institute for African Studies, University of Zambia by Manchester University Press.

de Wet, C. (2001, December 15). Economic development and population displacement: Can everybody win? *Economic & Political Weekly, XXXVI*(50), 4637–4646.

Fahim, H. M. (1981). *Dams, people and development: The Aswan High Dam case.* Amsterdam: Pergamon.

Falck, O., Heblich, S., & Link, S. (2011). The evils of forced migration: Do integration policies alleviate migrants' economic situations? *Discussion Paper, No. 5829*, IZA.

Fiala, N. (2009). The consequences of forced displacement in Northern Uganda. *HiCN Working Papers, No 65*, Households in Conflict Network.

Fiala, N. (2012). The economic consequences of forced displacement. *HiCN Working Papers, No. 137*, Households in Conflict Network.

Fiala, N. (2015). Economic consequences of forced displacement. *The Journal of Development Studies, 51*(10), 1275–1293.

Fuggle, R., & Smith, W. T. (2000). *Experience with dams in water and energy resource development in the People's Republic of China.* Cape Town: World Commission on Dams.

Hare, R. M., & Rawls, J. (1973). Rawls' Theory of Justice—II. *The Philosophical Quarterly, 23*(92), 241.

Howard, P. (1997). Development-induced displacement in Haiti. *Refuge, 16*(3), 4–11.

Ibáñez, A. M., & Vélez, C. E. (2008). Civil conflict and forced migration: The micro determinants and welfare losses of displacement in Colombia. *World Development, 36*(4), 659–676.

Kondylis, F. (2010). Conflict displacement and labor market outcomes in post-war Bosnia and Herzegovina. *Journal of Development Economics, 93*(2), 235–248.

La Rovere, E. L., & Mendes, F. E. (2000). *Tucurí Hydropower Complex Brazil.* Cape Town: World Commission on Dams.

Lam, L. M., & Paul, S. (2013). Displacement and erosion of informal risk-sharing: Evidence from Nepal. *World Development, 43*, 42–55.

Ortolano, L. (2000). *WCD case study: Grand Coulee Dam and Columbia Basin Project, USA.* Cape Town: World Commission on Dams.

Paul, S., & Sarma, V. (2017). Industrialisation-led displacement and long-term welfare: Evidence from West Bengal. *Oxford Development Studies, 45*(3), 240–259.

Puechguirbal, N. (2010). Women and displacement: Strength in adversity. Retrieved December 12, 2017, from https://bit.ly/2t39xuP.

Sarvimäki, M., Uusitalo, R., & Jäntti, M. (2009). Long-term effects of forced migration. *IZA Discussion Papers 4003*, Institute for the Study of Labor.

Scudder, T. (1996). Resettlement. In A. K. Biswas (Ed.), *Water resources: Environmental planning, management and development.* New York: McGraw-Hill.

United Nations, Department of Economic and Social Affairs, Population Division. (2017). *International Migration Report 2017: Highlights (ST/ESA/SER.A/404).*

Walicki, N., & Swain, M. (2016). Pushed aside: Displaced for "development" in India. July 2016. Retrieved February 21, 2019, from https://www.refworld.org/docid/57e920b24.html.

PART II

Regional Accounts: Socio-Economic Effects of Displacement

CHAPTER 4

The Long-Term Livelihood Effects of the Conservation-Led Displacement in Kanchanpur, Nepal

Lai Ming Lam, Basant Pant, and Vengadeshvaran Sarma

The Shuklaphanta Wildlife Reserve (National Park) Relocation Programme

The Shuklaphanta National Park (hereafter referred to as the "park") was established as a hunting reserve in 1969 and as a wildlife reserve in 1976 (Fig. 4.1). The government upgraded its status from wildlife reserve to national park in 2017 to generate revenue from wildlife tourism and promote indigenous communities natural resource use rights (GON 2017). The current aim of the park is to preserve its non-human habitants over an area of 155 km^2 of land (Upadhyaya and Yonzon 2003). Although the park is small, it is ecologically important as it is home to the world's largest

L. M. Lam
Osaka University, Suita, Japan

B. Pant
International Centre for Integrated Mountain Development (ICIMOD), Patan, Nepal

V. Sarma (✉)
University of Nottingham, Semenyih, Malaysia

N. Yoshino, S. Paul (eds.), *Land Acquisition in Asia*,
https://doi.org/10.1007/978-981-13-6455-6_4

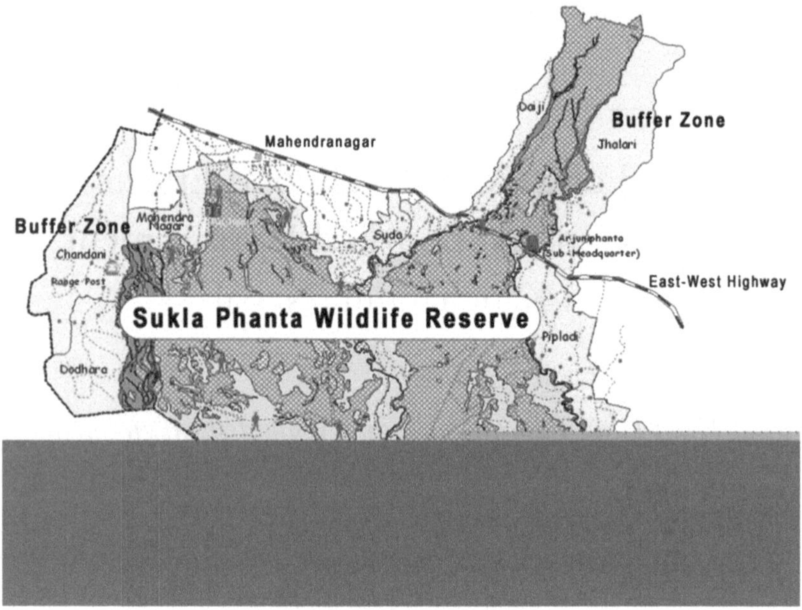

Fig. 4.1 Map of Shuklaphanta National Park and Buffer Zone, Nepal. Source: https://commons.wikimedia.org/wiki/File:Sukla-Phanta-WR%2Bbufferzone-map.jpg

population of swamp deer (approximately 2000), and its extensive grassland and swamps, along with the tropical and sub-tropical forests, support some endangered species such as tigers, elephants and rhinoceros. Moreover, 349 bird species, including 6 globally threatened species, have been recorded in the park (Upadhyaya and Yonzon 2003).

The park (then hunting reserve) first came about following the relocation of 47 families of the Haraiya Village and 15 families of the Rauteli Bichawa Village Development Committee (VDC) in 1974, this event affected more than 60 families. However, the ongoing development of new settlements adjoining it and activities such as logging, grazing and poaching further threatened the natural environment and wildlife habitats. The reserve was therefore considered too small for wildlife (Bhattarai 2001). Supported by global conservation institutions, in 1981, the Nepalese government announced that it was doubling the reserve's size to

305 km^2 (Bhattarai 2001). The result of this was that another 17 human settlements in five villages from ward 1 to 6 of the Rauteli Bichawa VDC were displaced from the extended area. Given the large-scale displacement caused by the programme, the royal directives emphasised the following (Bhattarai 2001):

1. All families displaced by the park extension programme were to be given the land they had lost.
2. Compensated land would be conducive for cultivation.
3. Resettlement areas needed to retain the social and cultural composition of displaced families.

After 20 years of work, the resettlement programme was finally completed in May 2002 with the support of the army. Within this time, 2249 households were relocated over seven locations that had a total area of 2108 hectares of cleared forest land. The displaced families were given similar-sized landholdings in relation to the property lost from the park extension works as seen through land registration records (Bhattarai 2001). There has, however, been no appeal process in place for families against the state's decision. Many illiterate indigenous farmers who had limited social connectivity to government officials and did not have land ownership documents often found the compensation unfair. For example, concerning one of our study sites, the largest resettlement area of Dhokka Block, at least ten households were categorised as illegal occupiers and did not receive any land compensation because they did not own official land documents.

The implementation of a resettlement programme was severely delayed due to households' poor and inaccurate record keeping, delayed relocation of households and rapid encroachment on resettlement sites, together with dramatic changes in the political environment and weak bureaucratic performance after the 1990s People's Movement.[1] In total, about 22 out of 27 commissions established were inaugurated following the 1990 movement (see, Bhattarai et al. 2017). Continuous reshuffling in government led to these commissions only having a short existence, lasting at the most for one year (Bhattarai et al. 2017). At the same time, people were given oral assurances of land compensation by the commissions/committees but the only gains made were political ones by: firstly, the political leaders who led the commissions and secondly, government employees who put some commissions/committees under political pressure, which led to the listing of land mafias (Bhattarai et al. 2017). Consequently, the distribution of land even to unlisted households took place due to some people's corrupt connections with commissioners (Bhattarai 2001). Finally, the park's resettlement programme took 20 years to complete and was only completed in May 2002 with the help of the army. All households inside the extension area were evicted.

The designation of the park is a typical fence-to-fence management model, which is now held by government authorities which include the Nepalese army and the park authority. Despite conservation organisations such as the UNDP, World Wildlife Fund and National Trust for Nature Conservation having been actively involved in park management and having implemented community development programmes in villages adjoining the park, one recent study documented that local participation in the park management was very low, and conflicts between conservation and local livelihoods continue to this day. One block of the national park was encroached on by 584 households in the Dhakka Block where at the time of relocation only 146 households resided, specifically, 46 households with registered land and 100 with unregistered land. This is the combined result of weak political governance, unplanned resettlement schemes and the district's population growth from 2001 to 2015 (Bhattarai et al. 2017). Huge forest areas were cleared off in the resettled sites in new Dhakka where the relocated people were resettled; this demonstrates the government's poor planning in terms of biodiversity conservation and its global conservation commitments. Moreover, it seems that the government does not have clear land use policies and development pathways for the district where the population is increasing. This situation gives a clear example of the current and potential conflict between conservationists and proponents of social and economic development. In this chapter, we therefore aim to assess the long-term impacts of conservation-induced displacement on local livelihoods and what the implications are for future conservation.

Research Methods

Our current study builds on the work of the authors over the past decade (see Pant 2007; Lam 2011; Lam and Paul 2013; Lam et al. 2016). The analyses conducted by Lam (2011) and Lam and Paul (2013) (hereafter referred to as the "2013 study") are based on an 18-month-long anthropological survey between 2004 and 2006 where 72 indigenous Rana

Tharu households living adjoining the park were interviewed. Pant's (2007) (hereafter referred to as the "2007 study") work is based on a quantitative and qualitative survey of 348 displaced households. Lam et al. (2016) (hereafter referred to as the "2016 study") used a *panel* sample of 56 households that was constructed using two rounds of surveys conducted in 2006 and 2013. To sum up, results from these previous studies indicate four key findings:

1. Post-displacement: Some of the households became landless, while others were compensated with land of poor quality and many people changed their livelihoods.
2. The cumulative effects of these were family partitioning, decline in their welfare and formation of new social safety nets.
3. In the long run, at least among one ethnic group, displacement seems to have had little impact on materialistic welfare.
4. In both the short and long term, however, there is evidence of declining emotional well-being and social cohesion among displaced households.

Pant (2007) and Lam (2011) indicate that after displacement, some of the displaced households became landless. Although land compensation was a key royal directive as part of the relocation programme, land compensation was often inadequate or absent. Part of this was related to the lack of land titles, as discussed earlier. Changes in livelihood structures coupled with inadequate harvest from the land also forced some households to sell their land to improve consumption patterns (Lam and Paul 2013). The authors further point out that lands provided as compensation in the resettlement areas were reportedly less fertile (Bhattarai et al. 2017) and therefore could not provide enough food for people's subsistence needs. The dearth of landholdings also led to family partitions as the risk-sharing based on familial landholdings became absent. These partitions first occurred between siblings (mostly married men) and later also affected parents and their sons. The rate of partitioning increased in tempo, to the extent that the authors discovered that one large, four-generation joint family broke down into six smaller families within ten years. Such partitioning is unusual given that Nepal is a patriarchal society, where male children traditionally continue to live with their parents after marriage, while daughters leave home to live with their in-laws. The study also notes that the relocation of households coupled with a deterioration of the patron-client system reinforced new social safety nets and led to large diversifications in people's livelihoods. Pant's personal on-site observations also indicate that the less fertile land did not provide sufficient food for resettled households, which then coupled with family fragmentation led to the selling of land that funded people's desire to go abroad. This resulted in degrading cultural intactness and consequently displaced households are more exposed to vulnerability.

While these initial studies focused on short-term effects, the 2016 study concentrates on the long-term impact of displacement using a panel sample of 56 indigenous Rana Tharu households (hereafter referred to as Ranas) from Rauteli Bichawa Village. Surprisingly, the study found that displaced Rana households are in fact better off than non-displaced Rana households in terms of food security and productivity. The authors, however, note that such material well-being is achieved at the expense of deteriorating social ties and unease about the future and increased stress on human capital. Some Rana families are abandoning agriculture activities altogether. The strain on human capital—via the need to find more work—also led to deteriorating familial and social cohesion among displaced households compared to non-displaced households. The study also reveals that some households who lost land and were among the worst affected moved out of the resettled areas. Although the study provides invaluable information so that we can better understand the long-term impacts of displacement, it has two major shortfalls: first, the relatively small sample and geographical concentration led to concerns of whether the findings can be generalised across Nepal let alone globally and second, it fails to identify heterogeneities among ethnic groups which may be important given the ethnic concentration of villages prior to displacement. In the other words, do all displaced households share the same experience as Ranas?

To address the above shortcomings in the 2016 study, we use data from the previous 2007 and 2013 studies to create a new panel dataset. The panels from both studies comprise two rounds of surveys carried out in 2006 and 2013. While the 2013 study's data comprise both displaced and non-displaced Rana households, the panel dataset built on the 2017 study consists only of displaced households yet includes all ethnicities. Pant's (2007) study involved 348 interviewed displaced households on the basis of three land compensation categories: first, obtaining an equal amount of compensation, that is, the total amount of land they had (full land compensation); second, receiving 0.34 hectares (10 kattha) of land; and third, getting 0.17 hectares (5 kattha) of land. However, due to data loss and attrition, we could only identify a list of 180 displaced households. These households were resettled in various villages, namely Jhalari, Laximpur, Rampur and Kalika. By tracing the displaced households to their original villages prior to displacement, we confirmed that the interviewed households in these two studies did not overlap. Using the two panels we address the following research questions:

1. Did all displaced Rana households share similar characteristics prior to and after displacement?
2. Do findings from the 2016 study hold if Rana households from the 2007 study are included?
3. Was there heterogeneity in the welfare of displaced households along ethnic lines?
4. Were non-Rana households better off or worse off compared to Rana displaced households over time?

We administered a second round of surveys in 2013. With some attribution bias, we successfully created a panel dataset encompassing 31 displaced Rana households in the 2013 study and 140 displaced households from all ethnicities documented in the 2007 study. This led to 171 displaced households, the details of which are summarised in Table 4.1. In this study, we could not establish a panel dataset on non-displaced households as the control group because non-displaced households were not included in the 2007 study. However, we include 25 revisited non-Rana households used in the 2016 study as a reference point.

In addition, two authors of this chapter, Lam and Pant, have conducted extensive research in the park area over a decade; their first-hand experi-

Table 4.1 Ethnic break-down of interviewed households

	2007 study (Pant 2007)	*2013 study (Lam and Paul 2013)*
Ranas	50	31
Dangauras	32	
BCNs (Brahmin, Chhetri and Newar)	54	
Dalits	4	
Total	*140*	*31*

ence provides pertinent insights into this topic which the quantitative data cannot inform us about, in particular the complex changes in people's livelihoods in the post-displacement period. Lam is trained as a cultural anthropologist and conducted ethnographic fieldwork with indigenous Rana communities in the park. Pant is a community inhabitant close to the park region and is a conservation management expert.

Supported by both quantitative and qualitative data, this study contributes to the literature in three substantial ways. First, to the best of our knowledge, this is the only study that uses a panel of all displaced ethnic groups related to the park relocation programme; the results provide us with a deeper understanding of the long-term welfare impact of displacement. Second, we are able to identify how ethnic and geographical localisations of displacement studies may affect the level of generalisation of findings. Third, the study evaluates heterogeneities in the effects of displacement and welfare along ethnic lines in Nepal. The latter is discussed in more detail in the following chapter.

Results

We use basic descriptive statistics, cross-tabulations and simple linear regressions to identify the heterogeneities in the effects of displacement across ethnic and geographic identities. The displaced Rana households in both panel datasets share similar livelihood trajectories despite their differences in terms of geographical locations and landholding sizes. Table 4.2 illustrates that displaced Rana households in Pant's panel are better off in 2013, a decade since their displacement. The improvement in their livelihood was even better than that of the displaced Rana households in the 2016 study.

According to Pant's panel dataset, except for canals, the displaced Ranas have better access to basic facilities such as water taps, primary schools and health centres. This may to some extent alleviate livelihood challenges caused by displacement and generate a more positive recovery environment for those displaced (see Table 4.3). The data show that these displaced Rana households owned landholdings half the size of those displaced Rana households in the 2016 study.

Furthermore, proportionally twice as many Rana households in Pant's panel complained of poor soil quality in comparison to the displaced Rana households by Lam et al. (2016). Apart from poor soil quality, they pointed out that poor irrigation and soil drought also caused poor yields

Table 4.2 Livelihood diversification by ethnic composition

		No. of livelihood options	*Agriculture: own farms (%)*	*Agriculture: others' farms (%)*	*Agriculture: livestock (%)*	*Service (%)*	*Businesses (%)*	*Migrants (%)*	*Others (%)*
Lam and Paul (2013)	Non-resettled Rana households (control group)	3.06	92	92	88	28	16	29	0
	Resettled Ranas	2.99	100	38	91	30	3	33	0
Pant (2007)	Resettled Ranas	3.16	94	84	96	45	11	29	0
	Resettled Dangauras	3.19	91	84	97	43	23	10	0
	Resettled Dalits	3.75	100	100	100	33	25	50	0
	Resettled BCNs	2.89	100	69	93	63	13	35	5

Note: BCNs refers to the combination of Brahmins, Chhetris and Newars

Table 4.3 Access to facilities, landholdings and food sufficiency

		Water tap (%)	*Primary school (%)*	*Health centre (%)*	*Canal (%)*	*Land size (katta)*	*No. of months' food available for*	*No. of months food available for per katta of land held*
Lam and Paul (2013)	Non-resettled Rana households (control group)	84	100	68	92	46.26	1.10	0.55
	Resettled Ranas	17	79	14	76	49.97	1.38	0.90
Pant (2007)	Resettled Ranas	88	98	69	4	25.29	1.46	1.63
	Resettled Dangauras	100	100	79	0	15.07	1.24	1.56
	Resettled Dalits	60	90	80	30	21.90	1.69	0.86
	Resettled BCNs	86	100	44	14	12.61	1.35	1.64

on their land resettlement. Despite all these complaints, food security and land productivity of these Rana households were surprisingly higher than both displaced Ranas and non-displaced Rana households in the 2016 study. Rana households in Pant's panel could be divided into two groups: small landholders and big landholders. Only one Rana household out of 50 revisited Rana households was landless while more than half of the displaced Rana households (63%) was small landholding owners who had 20 kattas of land or less and they were resettled in the Kalika village. Furthermore, one-third of the Rana households were resettled in Rampur where many previous big landlords including Ranas were accommodated. In contrast, it was a relatively homogenous scenario for the majority of Rana households who were surveyed in the 2016 study; they reported the existence of big landlords plus only a few nearly-landless Rana families. From the data and descriptive statistics, it is also evident that differences in the location of resettlement may have served an important role in shaping the livelihood changes among displaced households in the long term. In the case of the park resettlement programme, due to the shortage of cultivable land, most small landholding households were resettled in existing settlements like Kalika and Laxmipur villages, mixing with host families where basic facilities were readily available. In contrast, the big landlords were settled in the Rampur (Dhokka Block) area. According to interviews with these displaced households, they experienced difficulties in clearing up the virgin forestland for plantation and there was no health centre or schools in the area by the time they were relocated. We also observed that indigenous Rana households in Dhokka Block showed less interest in interacting with other ethnic groups.

If grow it yourself remains the major livelihood strategy for indigenous Rana households, it would be true that the more land you have, the better food security you will enjoy. The results implied major livelihood changes in the post-displacement context, that is, households no longer relied on their landholdings to provide their families with food. Families instead worked on others' farms to generate food for the family or attended to other ways of making a living. In this case, the food per capita per katta would in fact favour those families able to feed their families with the smallest possible land size. This may explain why relatively small landholding Rana households in Pant's panel experienced better food security than others.

This outcome was also consistent with the different livelihood strategies pursued by these two displaced Rana household groups. In Pant's panel, displaced Rana households had significantly higher levels of livelihood diversification than displaced Rana households in the 2016 study. For example, in order to overcome the problem of food deficiency, many Rana households relied on working for other people's agricultural land, doing business, or undertaking domestic and international labour migration to earn a living. It can be concluded based on our findings that displacement does not necessarily lead to materialistic poverty in the long run. In fact, on the contrary, it may have made displaced communities, like Ranas, more amenable to adaptation as shown in both the 2016 study and the current study.

The main question concerning the 2016 study was if their findings and livelihood trends identified above are unique to the Rana community or if the findings can be generalised to all other displaced household communities. Our panel dataset clearly indicates that displaced households from all ethnicities share similar experiences; to some extent, non-Rana households seemed to be better off in terms of land productivity after displacement compared to non-resettled Rana households. Surprisingly, the most disadvantaged group in Nepal, in terms of economics and social hierarchy, the Dalits, seemed to also have done better than non-resettled Rana households who were big landlords. The results reflect the dynamic livelihood changes in the post-displacement period which scholars and policymakers often tend to neglect because of the absence of any longitudinal study.

Livelihood is often approached as a materialistic subject as discussed in previous chapters. It should, however, be viewed as a multidimensional concept, particularly the role of social well-being—which plays an important role in people's overall livelihood. In our self-reported survey (Table 4.4), while displaced households have reported a general increase in food security, they reported *feeling* poor and unhappy compared to non-displaced households. These findings are similar to those reported in the 2016 study.

Discussion

First, the panel study helps us to better understand the fundamental nature of displacement which causes enormous destruction to every aspect of people's daily lives and the community: livelihood activities, social life and their psychological feelings regarding the future. Most policymakers recognise the economic losses caused by displacement and financial compensation schemes are subsequently designed to mitigate these "financial" losses. However, the social impact of displacement is often long lasting and may have little in common with improved material circumstances. According to our field observation in 2013, most displaced Rana households expressed their anger, anxiety and nostalgia for old community ties

Table 4.4 Emotional well-being—level of happiness

		Happy now? (1—Very unhappy, 5—Very happy)
Lam and Paul (2013)	Non-resettled Rana households (control group)	3.00
	Resettled Ranas	1.09
Pant (2007)	Resettled Ranas	1.78
	Resettled Dangauras	1.45
	Resettled Dalits	1.50
	Resettled BCNs	1.84

and homelands; the feelings of loss and unfairness have seldom faded away over time. For example, Madhan, a Rana participant interviewed in 2013, successfully improved his family's circumstances through out-migration, but he asserted that after the displacement, "everyone here is busy, working for *their* families and always thinking of money. Sharing is almost non-existent. We don't have the big Holi celebration as before." He further noted that while he thought their current livelihoods and financial conditions have improved since their displacement, he and others felt that remaining in their homeland would have made their lives better and more enriching.

Second, although the panel dataset shows that displaced Rana households in both the 2007 and 2016 studies are better off, we must be cautious in interpreting the livelihood status of the *missing* displaced Rana households. According to the 2016 study, 11 out of the 42 original displaced Rana households were not revisited as they could not be traced; this accounts for an attrition bias of almost 26%. These 11 missing displaced households, on average, according to our baseline study (Lam 2011), were economically worse off than the rest of the displaced Rana households. In Pant's panel dataset, 43 out of the 93 Rana households were missing, leading to an attrition bias of 46%. The total attribution bias for the whole panel study was 22%, and we could only revisit 140 out of the 180 displaced households. In fact, a mere 4 out of 18 Dalit families could be revisited in our second-round survey. Many small landholding farmers and landless households migrated from the resettled villages due to poverty. Therefore, the panel dataset may ignore the outcomes and well-being narratives of the economically poorest households in the first round of the surveys conducted by Lam (2011) and Pant (2007). That is, the displaced households who have land resources were most successful and tended on average to outperform the non-displaced Rana households but not the poorer land-deprived or marginal landholder households. This finding also reinforces the notion that historical and local contexts played very influential roles in determining people's welfare outcomes, systematically disadvantaging the poorest of the households.

Third, the results highlight the different levels of performance in terms of livelihoods along ethnic lines. A high proportion of displaced Rana households improved their livelihoods compared to others. It was due to them being indigenous inhabitants and many of them had previously been large landlords (*Jimidaris*). Although indigenous Ranas became increasingly marginalised after the influx of hill migrants, holding substantial land resources enabled the Rana elites to have more livelihood opportunities than other ethnic groups. For example, large landholder Rana households in Rampur were able to convert some of their land to plant cash crops such as sugarcane. They could also afford, in most instances, to send one of their family members to work abroad. The remittance helped them cope with the livelihood shocks and diversified their livelihood options. Some displaced Rana households started their own businesses or focused on long-term investments, even in human capital (their children's education). It was obvious that the new livelihood opportunities would not occur easily if households did not have enough basic livelihood assets, particularly land resources, and struggled for their everyday survival. It explains why later settlers such as the Chaudharys and Dangauras, Dalits and many missing landless displaced households, including some Rana households, did not share the same livelihood experience as displaced Ranas in the panel dataset. This result illustrates that displacement may not lead to long-term materialistic impoverishment for affected communities; however, it can further reinforce and enlarge the existing inequality gaps between the wealthy and poor household groups. Thus, involuntary displacement is an unjust act which often causes prolonged, negative impacts on vulnerable groups the most.

Conclusion

The novel panel dataset used in this study provides an in-depth perspective and helps us to understand the dynamic, comprehensive changes in the livelihood structures in the post-displacement period. While the indigenous Ranas and caste/ethnic groups have experienced material gains and improvements in their lives over a decade since displacement, this has, however, come at the expense of suffering anxiety, stress over livelihood difficulties and even loosening social ties. In other words, an overall deterioration in people's socio-cultural well-being is evident. Time has not helped displaced communities mitigate the sense of loss, pain and anger from displacement. The study raises a warning that displacement has caused prolonged and unlikely recoverable socio-psychological trauma for displaced communities. Land resources have become increasingly scarce due to population growth and the demands of economic development. The conflicts between wildlife preservation and human livelihoods are expected to become much tenser in the future. Can conservation-induced displacement be mitigated and justified? Unfortunately, the voices of affected inhabitants are often ignored in developing countries where most displacement has been carried out.

The establishment of a protected area has not only provided a comfort zone for nature and wildlife but it also has reshaped the socio-political and economic landscapes of the region. In the next chapter, we expand our study population to both displaced and non-displaced households adjoining the park from different ethnicities. What happens to people's livelihood changes after the creation and expansion of the park? Do caste/ethnic groups respond and act differently? All these scenarios offer us invaluable information to understand the process of displacement and social changes in a more holistic way.

Note

1. The 1990s People's Movement (Nepali: Jana Andolan) in Nepal was a multi-party movement. It brought an end to absolute monarchy and eliminated the Panchayat system. It marked the beginning of constitutional democracy. In 2006, following the restoration of absolute monarchy in Nepal, the Loktantra Andolan was launched, which once again illustrated various political parties' unity, leading some to brand it Jana Andolan II.

References

Bhattarai, A. M. (2001). *Displacement and rehabilitation in Nepal: Law, policy and practice*. New Delhi: Anmol Publications.

Bhattarai, S., Pant, B., & Timalsina, N. (2017). Conservation without participation: Detrimental effect of escaping people's participation in protected area management in Nepal. In S. A. Mukul & A. Z. M. M. Rashid (Eds.), *Protected areas: Policies, management and future directions* (Chap. 4, pp. 83–104). New York: Nova Science Pub. Inc.

Government of Nepal. (2017, February 20). Nepal Gazette—Chapter five. Ministry of Information and Communication, Department of Printing, Government of Nepal, Nepal.

Lam, L. M. (2011). Cultural perspectives of land and livelihoods: A case study of Royal Shuklaphanta Wildlife Reserve in the far-western part of Nepal. *Conservation and Society, 9*(4), 311–324.

Lam, L. M., & Paul, S. (2013, March). Displacement and erosion of informal risk-sharing: Evidence from Nepal. *World Development, 43*, 42–55.

Lam, L. M., Paul, S., & Sarma, V. (2016). Reversal of fortune? The long-term effect of conservation-led displacement in Nepal. *Oxford Journal of Development Studies, 44*(4), 401–419. https://doi.org/10.1080/13600818.2016.1149158.

Pant, B. (2007). *Impact of human relocation program on rural livelihood (A study from Shuklaphanta Wildlife Reserve, Nepal)*. A report submitted to Institute of Forestry, Pokhara, Nepal and The Rufford Maurice Laing Foundation, Rufford Small Grant for Nature Conservation (RSG), UK.

Upadhyaya, S. K., & Yonzon, P. (2003). *Societal diagnosis of the settlements in the Buffer Zone of the Royal Shuklaphanta Wildlife Reserve, Farwest Nepal.* Submitted to the Department of National Parks and Wildlife Conservation, Ministry of Forests and Soil Conservation of His Majesty's Government of Nepal.

CHAPTER 5

Attitudes Toward Land Acquisition in Indonesia

Naoyuki Yoshino, Rasyad A. Parinduri, and Yoko Oishi

INTRODUCTION

Inadequate Infrastructure Hinders Indonesia's Development

Despite the rapid expansion of the Indonesian economy, Indonesia has not fully developed its potential due to inadequate infrastructure. Although the country has the world's fourth-largest population, its economy continues to rely on the export of primary industry products such as coal, natural gas, and palm oil. The lack of adequate infrastructure has left the business environment unattractive. According to the Global Competitiveness Report 2016–2017, Indonesia ranks 41st out of 138 economies (World Economic Forum 2016). One of the reasons for this low competitiveness is the high cost of logistics. The World Bank reports that Indonesia dropped in the ranking of the Logistics Performance Index from 53rd in 2014 to 63rd in 2016, out of 160

N. Yoshino • Y. Oishi (✉)
Asian Development Bank Institute (ADBI), Tokyo, Japan
e-mail: nyoshino@adbi.org; yoishi@adbi.org

R. A. Parinduri
University of Nottingham, Malaysia Campus, Selangor, Malaysia
e-mail: Rasyad.Parinduri@nottingham.edu.my

N. Yoshino, S. Paul (eds.), *Land Acquisition in Asia*,
https://doi.org/10.1007/978-981-13-6455-6_5

economies in total. Its ranking is lower than other member countries of the Association of Southeast Asian Nations (ASEAN).[1] The data published by the Ministry of Trade (*Kementerian Perdagangan*) (2016) show that, on average, about 26.65 percent of GDP during 2004–2011 accounts for the cost of logistics in Indonesia. The Indonesian Logistics and Forwarders Association (*Asosiasi Logistik dan Forwarder Indonesia/* ALFI) reported that the cost of logistics declined to 23.5 percent in 2017, but it is still higher than Singapore's 8.1 percent and Malaysia's 13 percent (ADB 2017). In addition, despite the abundant energy resources, Indonesia has not succeeded in the provision of stable electricity, especially outside major cities. Missing links in the road network and the unstable power supply also burden the life of the general population. They cause inequality in opportunity and weaken social security and food and nutrition security. It is difficult for children to attend school and for the population to access health-care facilities without a usable road network. Moreover, lack of well-developed road networks creates large price differences across the archipelago. Food prices are not an exception. For example, rice is far more expensive in eastern Indonesia than on the islands of Java or Sumatra. Even in Indonesia's capital, Jakarta, domestically produced fruit are much more expensive than the ones imported from abroad. Furthermore, in the rainy season, fragile infrastructure is devastated by floods but also exacerbates the damage caused by floods. And this damage leads to supply shortages and inflation in different regions. This is despite being the world's largest archipelago and having large aquatic and marine resources. Indonesia's seafood business is far from being a flourishing industry due to expensive logistics and lack of cold supply chain. According to World Bank (2016) estimates, in Indonesia, the households at the low end of the income scale spend as much as 61 percent of their income on food and about one-third of their income on rice alone. Obviously, this situation is not favorable both to consumers and to producers. Securing the right quantity of and quality of infrastructures is in alignment with agriculture policy. According to the Strategic Plan of the Indonesian Ministry of Agriculture 2015–2019, the government is carrying out comprehensive missions to improve the situation. The list includes (1) food sovereignty, (2) a sustainable agriculture-bioindustry system, and (3) farmers' welfare.[2]

Land Acquisition Is the Primary Obstacle in Infrastructure Development

Aware of the importance of infrastructure development, the Indonesian government has accelerated infrastructure development. The national medium-term development plan (RPJMN 2015–2019) lists 245 infrastructure projects, including toll roads, railroads, harbors, airports, refineries, power plants, gas distribution, and irrigation. As part of this development plan, the Widodo administration aims to decrease the ratio of logistical costs to GDP from 23.5 percent to 19.5 percent. The government has increased the budget available for infrastructure development[3] and has mobilized private investment and soft loans—mainly from China and Japan—to push the plan forward.

However, the projects continuously face a common obstacle: land transfer. The Indonesian coordinating minister for Economic Affairs said, "44 percent of the problems in implementing national strategic projects relate to land acquisition".[4] Although government and developers are figuring out ways to finance projects, they suffer from the sluggish land transfer, as infrastructure projects are vulnerable to delays. Land acquisition is a pressing issue for the Indonesian government.

Local communities also suffer from land acquisition. For example, a land dispute on the construction of the new Yogyakarta airport has been continuing since 2011. In this dispute, local authorities and developers cut the supply of electricity in November 2017 to force residents to relocate (Muryanto 2018). However, about 300 residents still refused to relocate and remained on the site in April 2018. In a more recent case, the Jakarta-Bandung high-speed rail-line project scheduled for completion in 2019 has been delayed due to land acquisition problems.

The reasons for the sluggish land acquisition process are very broad and include political, economic, and social factors. Political factors include the elected president's hesitancy for executing expropriation power,[5] unaligned incentives and authoritative power of the institutional arrangements within central and local governments,[6] and the people's suspicion of the corruption behind the public interest rhetoric (O'Rourke and Milne 2010; The Jakarta Post 2011). Economic and social concerns vary from low compensation to cultural values surrounding inherited land and traditional unwritten common law (*adad*).

Structure of the Chapter

The government's endeavor to smoothen the process of land acquisition for infrastructure projects, however, still requires political commitment and support from the landholders. This research investigates the relationship between landowners' willingness to lease land for infrastructure projects and the characteristics of the agricultural land, household head, and village. The investigation of the determinants of leasing will provide insights for land planning and the negotiation process. Also, we examine the lease and its characteristics. Ghatak and Ghosh (2011) claim that the calculation formula for compensation must reflect the dispossessed owners' valuation of their assets. However, during a negotiation, the landlord has an incentive to exaggerate the value of the land.

This chapter is organized as follows. The section "Background" reviews the legal framework regarding the procedures for land transfer and its amendment, the section "Literature" reviews related literature (mainly from Asian countries), the section "Methodology and Data" introduces our methodology and data, and the section "Results" presents the empirical analysis, and Conclusion offers concluding remarks of the chapter and provides policy implications.

Background

Land Acquisition Law

To address the issue of slow land acquisition, the government amended the Land Acquisition Law in 2012 (Law No. 2/2012). This law deals with the revocation of land rights to serve the public interest. Notable changes are consolidation of the land acquisition process under the umbrella of the National Land Agency (*Budan Pertanahan Nasional*/BPN) (Article 27)[7]; appraisal conducted using an appraiser accredited by the National Land Agency (BPN) (Article 31); the value of "other losses that can be assessed" included within the compensation (Article 33); and placing time limits on each procedural phase (Article 37–38). After promulgation, the government further amended the law by publishing several presidential decrees[8] and technical guidelines[9] in the form of a decree of the head of National Land Agency (BPN). It is strengthened by other institutional frameworks and related presidential decrees.[10]

Several scholars assessed the reformed legal framework positively, arguing that it speeded up the land acquisition process. However, others—especially private companies—criticize that the continuous amendment and complex nature of the applied law (which depends on the phase of the project) bring uncertainty and confusion.

Regarding compensation, the Land Acquisition Law (2012) was amended to improve the safeguards for citizens and land evaluators were required to consider the immaterial value of land. Before the amendment, compensation was based on the tax assessed value (*Nilai Jual Objek Pajak*/NJOP) which is far lower than the potential market value (World Bank 2007). In addition, in Indonesian Land Law, the release *pelepasan* or surrender *penyerahan* of land rights is considered voluntary, and compensation has traditionally covered only damage (Davidson 2016). Therefore, compensation is traditionally understood as merely a "damage replacement"—commonly called in Indonesia *ganti rugi*—and therefore it is comparatively low. In contrast, the amended law requires the use of third-party land appraisers and mandates that the non-physical value be considered in the appraisals. Under the amended law, compensation contributes in reimbursing the landowners for the incurred loss to some extent. However, there still is criticism that compensation is flawed, as it is based on the land market value from a few years prior (when the gubernatorial location permits were issued) (presidential decree No. 30/2015). This causes lower compensation than current market value.

Appraisals for Compensation

Under the amended law, the compensation amount is appraised as the sum of the market value of land, space above and underground, buildings, plants, objects related to land, and/or other losses that can be assessed (Law No. 2/2012: Article 33). The last item "other losses" is a vague notion, but the law's elucidation defines it as "non-physical damage *disetarakan* that can be equated with a monetary value, for example, damages due to the loss of business or work, moving costs, costs associated with changing professions and with the loss of value of residual property". In practice, to implement the legal mandates of the amended Land Acquisition Law, the Indonesian Society of Appraisers (*Masyarakat Profesi Penilai Indonesia*/MAPPI) has established a regulatory framework for appraisal of compensation: "Indonesian Valuation Standard (SPI 306), Valuation in the Context of Land Acquisition for Development for Public Interest[11]". This standard emphasizes fair replacement value *nilai penggantian wajar*, full compensation to the owner for losses, with special attention to elements of non-physical losses caused by relinquishing property ownership. The standard includes a compensation valuation method for physical and non-physical losses that include a premium rate for emotional loss *solatium* amounting to 10–30 percent of the total rate for physical loss, lost jobs or business or profession change, transaction cost, interest, loss of remaining land, and other physical losses (Asian Development Bank 2017).

Procedure After the Appraisals

The government clarifies the procedure after an appraisal and sets the time for each procedure to shorten the process in the amended law. However, the procedure is criticized for being less favorable to landowners, as it is a one-sided decision. The appraisal result is reported to the National Land Agency (BPN). The agency should deliberate with the entitled party within 30 working days after the report (Law No. 2/2012: Article 37). In the case that the compensation amount is not acceptable to the landowner, the amended law allows the entitled party to submit an objection to the district court within 14 working days from the determination of compensation (Law No. 2/2012: Article 38). The district court has no more than 30 working days to decide the form and amount of compensation. The plaintiff may appeal within 14 working days to the Supreme Court. The Supreme Court provides the final decision within 30 working days. The decision made by the Supreme Court is definitive and binding.

The government grants landowners the right to file a lawsuit. However, for landowners, especially the ones who are not familiar with the legal process, it could be a heavy burden to file a case. Another bottleneck is that developers are not required to disclose the entire information on the compensation; the amended law only requires developers to disclose certain information from their land acquisition plan (Law No. 2/2012: Articles 16, 18–21), but the proposed options for compensation and resettlement are not required to be disclosed (Aziz 2017: The Conversation). However, this information is important for the affected parties to decide whether or not to accept the conditions offered. It also plays a crucial role in lawsuits.

Moreover, even if landowners bring cases to the court, there remain issues in establishing a mutually positive settlement for smooth land acquisition. Aziz (2017) collected data from the Supreme Court and found that many citizens lost their case and failed to receive proper compensation. He also pointed out that in cases where citizens won in the district court, the Supreme Court would overturn the decision. In the case of construction of the new Yogyakarta airport, for example, there were 222 cases handled by the Wates District Court (Aziz 2017). According to Aziz's research, the reason farmers lost in court was because they could not prove that the compensation they demanded was reasonable while the executors would call a land appraisal officer as their witness to testify in court and support their proposed amount. For a fair evaluation of the property and legal protection of the landlord's rights, it is crucial to develop fair and nationally accepted standards for compulsory acquisition procedures of infrastructure construction, the assessment of willingness for the transfer, and optimal compensation.

Literature

Willingness for Relocation and Lease

Although the determinants of the willingness for reallocation may depend on regional or country characteristics, few studies on the topic have been conducted in Asian countries. A study in the People's Republic of China (PRC) reveals that people are concerned most with compensation in the pre-eminent domain. A total of 86 percent of the respondents showed concerns about compensation, followed by housing (66 percent), education (63 percent), land (52 percent), and employment (52 percent), among others (Heming and Rees 2000). Regarding the characteristics of the landowner and willingness to relocate, Fang et al. (2016) report on the relationship between age and willingness from their study on the socio-economic characteristics of the successful and failed cases of rural construction land consolidation in the PRC. Their study found that, the older the household head is, the greater the reluctance to relocation. However, when comparing the groups "over 70" and "between 60 and 70", the oldest age group has a higher probability to agree to relocation. The authors explain that the amount of compensation provided is seen as more than they can use within their remaining lifetime by the former group, while the latter group thinks it is not enough. Besides age, Masterson (2007) found that the sex of the household head matters to land market participation, as it is highly related to credit availability. This is based on evidence from a regression analysis conducted in Paraguay in 1990. This research assumes that the determinants of land lease include characteristics of the household, their production, and location. Regarding household characteristics, the variable includes the number of adult household members, the amount of land owned, and the age of the household head. Further, the regression includes variables capturing whether they use credit to purchase inputs and capital equipment, productive assets, land tenure, and regional dummies for the location of the land.

Compensation Calculation

According to the Indonesia Valuation Standard (SPI 306), a land appraisal is conducted by both the market approach and the income approach (MAPPI). A land price is generally difficult to obtain in rural areas, due to the limited number of land transactions occurring through the market. Yet, it is widely used as an anchor for calculating compensation. For example, in India, a legally specified compensation formula is based on market price or a market price plus mark-up and stipends (Land Acquisition, Rehabilitation and Resettlement Bill/LARR 2011). However, Ghatak and Ghosh (2011) claim that the market price of rural land is not an adequate benchmark for compensation since the rural land market is not as well developed as that in urban areas.

For the price of land that is planned to be developed, some researches argue that in addition to formal market value, the potential development should be included (Hushak and Sadr 1979; Chicoine 1981; Shonkwiler and Reynolds 1986; Elad et al. 1994; Mendelsohn et al. 1994; Vitaliano and Hill 1994; Shi et al. 1997; Plantinga and Miller 2001; Hardie et al. 2001). Since one cannot take the future value into account unequivocally, Capozza and Helsley (1989) propose an urban spatial model with stochastic development rents. The feature of this model is a bid rent for developed land that is a function of the distance from the center of economic activity (Mills 1981; Capozza and Helsley 1989).

Others estimate the future value by conducting empirical analyses using income, population, and accessibility variables as determinants of the increased farmland prices due to development. In addition, other literature suggests including option values, since the expected benefit from development involves uncertainty. For the theoretical and empirical studies on the effect of option values on land development, see, for example, Titman (1985), Quigg (1993), Capozza and Sick (1994), Tegene et al. (1999), Capozza and Li (2002), and Turvey (2002).

All the theoretical and empirical studies require further consideration as fair compensation, which could be supported by unobserved factors. For instance, Ghatak and Ghosh (2011) claim that the value for the landlord depends not on tangible attributes, which can be observed by evaluating officials, but on subjective elements. He elaborates that even the flow of crop output depends on subjective factors such as the endowment of skill, knowledge, capital, farming assets or technologies, market access, and access to an alternative method of earning a livelihood. In addition, the land has psychological values or values as self-helped social securities "collateral for lands, assured source of employment for family labor, insurance against food price fluctuations via self-consumption and even social prestige associated with landownership" (Ghatak and Ghosh 2011).

Means of Compensation

The means of compensation also varies. For example, remuneration methods include lump sum cash payments, employment guarantees, annuities, company shares, land for land, a share of appreciated land value after resale, replacement of lost homestead, and so on. Compensation in cash is criticized for replacing a farmer's familiar assets, that is, farmland, with unfamiliar cash and increasing vulnerability by depriving the farmer of the opportunity to utilize asset-specific skills. Cash also incurs self-control problems associated with liquid assets (Banerji and Ghatak 2009). Regardless of these shortcomings, monetary compensation is widely accepted. The Land Acquisition for Public Interest Development allows that the distribution of compensation be granted in the form of money, land replacement, resettlement, share ownership, or other forms approved by both parties (Law No. 2/2012: Article 36).

Methodology and Data

We examine the determinants of landlords' willingness to rent out their land using the following regression:

$$y_{ijk} = \alpha + X_{ijk}\beta + W_i\gamma + Z_{ij}\eta + \varepsilon_{ij}$$

where y_{ijk} is an indicator of whether household head i who lives in village j is willing to rent out his or her plot k for infrastructure projects (if private buyers are available), X is a vector of plot k's characteristics, W is a vector of household head i's characteristics, Z is a vector of village j's characteristics where person i lives, and ε_{ij} is the error term. In selected specifications, we also include householder village fixed effects to control for unobserved time-invariant households or village characteristics (in the former case, we would drop W from the regression, in the latter W and Z). In addition to landlords' willingness to rent out their land, we also use another dependent variable—how much they like to receive per hectare as rent (if they like to relocate but do not want to sell their land).

We get the data from a survey we conducted on farm households in 130 villages in 26 sub-districts of 13 districts in Central and East Java in 2018. In each village, the survey selected 14 households. Therefore, we have 1820 households in total. Among those households, 652 of them are owners of 2003 plots of land. We interviewed households to collect detailed information about the plot and the attitude toward leasing it. The plot characteristics were reported from landowners as for their main two plots used for rice cultivation. In addition, we presented a hypothetical situation to each landlord regarding those two plots: If the government wants to build infrastructure on their land, and private buyers of their land are available, would they be willing to rent out land to the government? We also ask them to consider this: How much would they like to receive as rent if they like to relocate but do not want to sell their land? Additionally, we interviewed village officials of sampled 130 villages to collect information on general village characteristics.

We have three sets of variables in the survey which we consider as determinants: plot, household, and village characteristics. Plot characteristics include whether households self-cultivate the plot, whether they obtained the plot through inheritance, whether the plot is a wetland,[12] whether it is located in the village, the number of years they have owned the plot, its area, and the ratio between the price they ask for the plot if they have to sell it and its current price. Household characteristics include the household heads' age, gender, marital status, and educational attainment; whether he or she worked in agricultural sectors; whether he or she has worked in non-agricultural sectors; and the number of household members. Village characteristics include the villages' population, size of land, and size of non-agricultural land; the distance between the village and nearest bus terminals; whether the village is accessible by asphalt or cement road; and whether the village is in a rural area.

The summary statistics we present in Table 5.1 suggests some determinants of landlords' willingness to rent out their land. Landlords seem to be more likely to be willing to rent out their land if they do not self-cultivate the plots or obtain it through inheritance, if the plot is not located in their villages, and if they have not owned the plots for long. Younger, male, married, and better-educated household heads are more willing to rent out their plots. Bigger households are more willing, so are households whose heads work in non-agricultural sectors. Landlords who live in bigger villages and in villages whose area of non-agricultural land is large are more willing to rent out their plots.

The table also suggests that landlords who are willing to rent out their plots of land demand higher rents, ask for higher prices if they decide to sell their land, and quote higher market prices for their plots (the top

Table 5.1 Summary statistics

Variables	*Would you be willing to rent your land out to the government for infrastructure-related projects if private buyers are available?*		*Would you be willing to rent your land out to the government for infrastructure-related projects if private buyers are not available?*	
	No	*Yes*	*No*	*Yes*
	(1)	*(2)*	*(3)*	*(4)*
A. Plot characteristics				
Self-cultivated	0.60	0.56	0.60	0.62
	(0.49)	(0.50)	(0.49)	(0.49)
Obtained through inheritance	0.59	0.56	0.58	0.65
	(0.49)	(0.50)	(0.49)	(0.48)
Wetland	0.81	0.84	0.81	0.79
	(0.39)	(0.37)	(0.39)	(0.41)
Located in the village	0.92	0.87	0.92	0.93
	(0.27)	(0.33)	(0.27)	(0.25)
The number of years of owning the land	20.52	18.77	20.53	20.40
	(15.96)	(15.06)	(15.85)	(16.97)
Area (hectares)	0.45	0.96	0.44	0.47
	(2.04)	(3.94)	(2.07)	(1.70)
Rents and prices				
Expected rent (Rp million)	6.17	10.34	6.07	7.00
	(19.40)	(28.36)	(19.94)	(14.02)
Expected price (Rp million)	334.41	458.70	335.27	327.08
	(637.75)	(1018.93)	(617.81)	(792.02)
Current price (Rp million)	276.60	390.73	279.70	249.98
	(549.65)	(991.45)	(552.99)	(521.83)
The ratio of expected price and current price	1.36	2.75	1.35	1.43
	(1.33)	(40.83)	(1.38)	(0.77)
The difference between expected and current prices (Rp million)	57.81	67.97	55.57	77.10
	(258.87)	(260.93)	(244.40)	(360.95)
B. Household-head characteristics				
Age	57.07	54.04	57.52	53.27
	(11.94)	(11.41)	(11.85)	(12.10)
Male	0.86	0.94	0.85	0.91
	(0.35)	(0.24)	(0.35)	(0.29)
Married	0.84	0.91	0.84	0.86
	(0.37)	(0.29)	(0.37)	(0.35)

(*continued*)

Table 5.1 (continued)

Variables	*Would you be willing to rent your land out to the government for infrastructure-related projects if private buyers are available?*		*Would you be willing to rent your land out to the government for infrastructure-related projects if private buyers are not available?*	
	No	*Yes*	*No*	*Yes*
	(1)	*(2)*	*(3)*	*(4)*
Educational attainment	8.27 (4.59)	10.45 (4.50)	8.11 (4.62)	9.66 (4.06)
Worked in the agricultural sector	0.57 (0.50)	0.46 (0.50)	0.57 (0.50)	0.57 (0.50)
Worked in the non-agricultural sector	0.30 (0.46)	0.44 (0.50)	0.30 (0.46)	0.31 (0.47)
Household size	3.57 (1.50)	3.97 (1.57)	3.55 (1.50)	3.78 (1.49)
C. Village characteristics				
Rural area	0.80 (0.40)	0.80 (0.40)	0.80 (0.40)	0.82 (0.39)
Population (thousand)	3.73 (2.31)	3.81 (2.64)	3.75 (2.31)	3.57 (2.36)
Size (thousand hectares)	0.89 (7.02)	1.51 (9.94)	0.77 (6.28)	1.92 (11.59)
Size of non-agricultural land (thousand hectares)	0.15 (0.84)	0.24 (1.19)	0.14 (0.75)	0.27 (1.39)
Distance to the nearest bus terminal (kms)	6.99 (7.36)	6.86 (7.28)	6.74 (7.13)	9.15 (8.81)
Time it takes to the nearest bus terminal (hours)	0.25 (0.24)	0.23 (0.21)	0.25 (0.24)	0.27 (0.24)
Distance to the district capital (kms)	20.90 (12.07)	21.17 (12.85)	20.74 (11.93)	22.28 (13.23)
Time it takes to the district capital (hours)	0.62 (0.35)	0.64 (0.35)	0.62 (0.35)	0.63 (0.31)
Asphalt or cement road	0.92 (0.27)	0.88 (0.32)	0.92 (0.27)	0.89 (0.32)

Note: The number in each cell is the mean, those in parentheses are standard deviations. The numbers of observations for columns 1, 2, 3, and 4 are about, respectively, 1100, 900, 1000, and 100 rows). Landlords who are willing to rent out their land demand 68 percent higher rents per hectare, ask for 37 percent higher prices if they decide to sell their plots, and quote 41 percent higher market prices.

Correlation coefficients between landlords' willingness to rent out their land and the determinants are typically low: Some of the highest are educational attainment (0.25), working in non-agricultural sectors (0.16), age (−0.14), male (0.13), and married (0.12).

The same applies to the correlation between rents that landlords demanded and the determinants: Some of the highest are area (0.20), household size (0.09), and age (−0.07). (We do not present the tables of correlation coefficients for the sake of brevity.) Some scatter plots in Fig. 5.1 illustrate the correlation between rents demanded and some of the continuous determinants: We see that rents demanded are positively correlated with ask prices and negatively correlated with land area. For other variables, the relationship is less obvious.

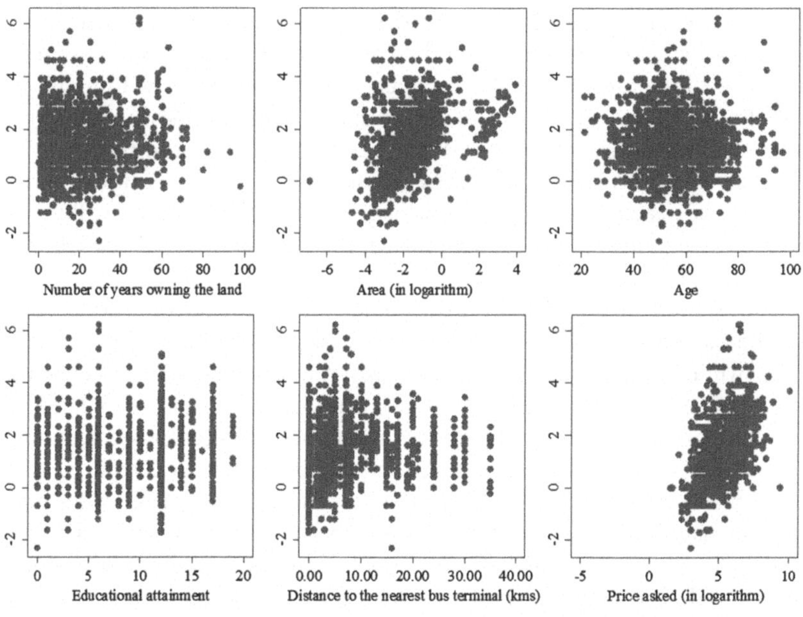

Fig. 5.1 Scatter plots of rents demanded by landlords for infrastructure projects and some determinants

Results

Willingness to Rent Out Land

We find that landlords are less willing to rent out their land if they self-cultivate the plots, if plots have been obtained through inheritance, if plots have been owned for a longer period of time, if the land area is small (particularly in the case of non-wetlands and plots located in their villages), and if the ask price demanded is higher than current prices (the top part of Table 5.2). Only the estimates of the coefficients of *self-cultivated*, *in the village*, the *number of years of owning the land*, and *area* are statistically significant in some specifications, however. The estimates are similar irrespective of whether we add household or village characteristics as independent variables (columns 1–3). Landlords who self-cultivated and inherited plots are about 5 and 3 percent less likely to be willing to rent out their plots or 5 and 8 percent in the case of non-wetlands and plots located in the village. One additional hectare of plot area is associated with about 10 percent higher likelihood of willingness to rent out plots. A one percent increase in the ask price above market price is associated with 0.4 percent lower likelihood of willingness to rent out the land. When we introduce village fixed effects (we rely more on within-village variations), none of the estimates is statistically significant, although almost all have the same signs. When we include household fixed effects (we rely more on within variations across plots of the same household), the sign of *wetlands*' estimates turns negative (landlords are less willing to rent out wetlands) and the magnitude of the *ratio of expected and current prices* quadruples.

Household heads who are male, older, more educated, who do not work in agricultural sectors, and who have bigger families are more likely to be willing to rent out their land. However, married household heads are less willing to rent out their land (the middle part of Table 5.2). The estimates are statistically significant across the different specifications, except for those of *age*, *married*, working in agricultural sectors, and working in non-agricultural sectors. They also have the same signs and similar magnitude. Landlords who have one more year of schooling, are male, are unmarried, and have one more household member are associated with 1–3 percent higher likelihood to be willing to rent out their plots.

Table 5.2 Determinants of willingness to rent out land if there are private buyers

Dependent variable: Willing to rent your land out to the government for infrastructure-related projects if private buyers are available					
	(1)	*(2)*	*(3)*	*(4)*	*(5)*
A. Plot characteristics					
Self-cultivated	−0.059*	−0.051*	−0.047	−0.054	0.001
	(0.023)	(0.026)	(0.026)	(0.030)	(0.029)
Obtained through inheritance	−0.010	−0.035	−0.033	−0.024	0.001
	(0.023)	(0.023)	(0.023)	(0.028)	(0.028)
Wetland	0.049	0.051	0.053	0.026	−0.160*
	(0.030)	(0.029)	(0.030)	(0.061)	(0.064)
Located in the village	−0.115**	−0.082*	−0.081*	−0.047	−0.037
	(0.038)	(0.036)	(0.036)	(0.050)	(0.033)
The number of years of owning the land	−0.002**	0.000	0.000	−0.001	−0.001
	(0.001)	(0.001)	(0.001)	(0.001)	(0.001)
Area (hectares)	0.013**	0.013**	0.012**	0.005	−0.000
	(0.003)	(0.003)	(0.003)	(0.004)	(0.001)
The ratio of expected and current prices	0.004**	0.004**	0.004**	0.005**	0.003**
	(0.000)	(0.000)	(0.000)	(0.001)	(0.000)
B. Household-head characteristics					
Age		−0.001	−0.001	0.001	
		(0.001)	(0.001)	(0.002)	
Male		0.182**	0.174**	0.155	
		(0.059)	(0.059)	(0.082)	
Married		−0.076	−0.072	−0.086	
		(0.054)	(0.054)	(0.078)	
Educational attainment		0.020**	0.020**	0.022**	
		(0.003)	(0.003)	(0.005)	
Worked in the agricultural sector		−0.011	−0.026	−0.077	
		(0.041)	(0.042)	(0.072)	
Worked in the non-agricultural sector		0.046	0.027	−0.001	
		(0.043)	(0.043)	(0.076)	
Household size		0.031**	0.032**	0.030**	
		(0.008)	(0.008)	(0.010)	
C. Village characteristics					
Rural area			0.008		
			(0.033)		
Population (thousand)			−0.005		
			(0.006)		
Size (thousand hectares)			−0.046*		
			(0.019)		

(*continued*)

Table 5.2 (continued)

Dependent variable: Willing to rent your land out to the government for infrastructure-related projects if private buyers are available					
	(1)	*(2)*	*(3)*	*(4)*	*(5)*
Size of non-agricultural land (thousand hectares)			0.417**		
			(0.156)		
Distance to the nearest bus terminal (kms)			0.002		
			(0.002)		
Time it takes to the nearest bus terminal (hours)			−0.171*		
			(0.067)		
Distance to the district capital (kms)			−0.000		
			(0.001)		
Time it takes to the district capital (hours)			0.014		
			(0.045)		
Asphalt or cement road			−0.058		
			(0.042)		
Fixed effects					
Village fixed effects				Yes	
Household fixed effects					Yes
Adjusted R-squared	0.017	0.086	0.093	0.087	0.014
The number of observations	1917	1912	1912	1912	1917

Note: The figures in parentheses are robust standard errors. * and ** indicate statistical significance at levels of 5 percent and 1 percent, respectively

Among the village characteristics that we include as independent variables, we find landlords are more likely to rent out their land if the villages they live in have larger areas of non-agricultural land, have a smaller size, and are closer to sub-district capitals where bus terminals are usually located (the bottom part of Table 5.2). Only the estimates of the coefficients of the area of non-agricultural land and village size are statistically significant: A decrease in village size, and an increase in the size of non-agricultural land, by one hectare, increase the likelihood of being rented out by 5 and 45 percent, respectively. The estimates of ***population*** and ***rural area*** coefficients are small, particularly the coefficient on ***rural areas*** (plots in rural areas are associated with only 0.1 percent more likelihood of willing to be rented out by the landlords).

Rents Demanded

We do the same analysis with rents demanded by landlords who are willing to rent out their plots for infrastructure projects, but the results are not as clear as in the analysis on the determinants of willingness to rent out plots: The estimates are all statistically insignificant except the estimates of the coefficients of two household heads' characteristics, age, and educational attainment (Table 5.3). We find landlords demand higher rents if they obtain their plots through inheritance, work in agricultural sectors, and have a large area of land, in particular, non-wetlands and plots located in their villages. Among village characteristics, we find that rents in bigger villages are higher but the bigger the size of non-agricultural land, the lower rents will be. We also find that younger and more educated landlords demand more rents.

Conclusion

Designing a fair and socially accepted land acquisition scheme is crucial for expanding and improving infrastructure in Indonesia. As outlined in "Background", the government has sought to shorten the process of land acquisition by clarifying the procedure and setting time limits for each step. At the longest, the process via court takes about 4.5 months under the amended Land Acquisition Law (from the time appraisal is reported to the National Land Agency (BPN) to the final decision made by the Supreme Court).

However, lawsuits put an undue burden on landowners. Instead, offering acceptable compensation and avoiding legal dispute facilitate more timely and smoother land transfers. To this end, it is paramount that the process of compensation be designed in a way that guarantees fair and mutually beneficial results. Transparency is key to avoiding moral hazards.

Our research shows that landowners' willingness to lease and their expected rental price depend on whether the plots are wetlands, whether they are located in their local villages, and on the size of the plot. Currently, the premium for non-physical loss is calculated based on the physical value of the relinquished property. If the process of appraisal was instead designed in such a way that the above factors are taken into account, it would encourage landowners to accept the proposed compensation with a greater sense of fairness. Such an approach would allow the government

Table 5.3 Rents landowners ask if they do not want to sell the land

Dependent variable: How much would you like to receive per hectare as rent if you like to relocate but do not want to sell your land?

	(1)	(2)	(3)	(4)	(5)
A. Plot characteristics					
Self-cultivated	−0.184**	−0.110*	−0.088	−0.123*	0.009
	(0.049)	(0.053)	(0.051)	(0.060)	(0.070)
Obtained through inheritance	0.020	0.010	0.025	0.072	0.060
	(0.048)	(0.048)	(0.047)	(0.046)	(0.048)
Wetlands	0.095	0.119	0.193**	0.120	0.031
	(0.069)	(0.069)	(0.070)	(0.103)	(0.119)
Located in the village	−0.254**	−0.244**	−0.209*	−0.163	−0.214**
	(0.083)	(0.082)	(0.081)	(0.088)	(0.080)
The number of years of owning the land	−0.000	0.002	0.001	−0.001	0.003
	(0.002)	(0.002)	(0.002)	(0.003)	(0.003)
Area (hectares)	0.067**	0.068**	0.037**	0.047**	0.050**
	(0.005)	(0.005)	(0.005)	(0.007)	(0.014)
The ratio of expected and current prices	−0.004*	−0.004*	−0.003	0.002	−0.001
	(0.002)	(0.002)	(0.002)	(0.003)	(0.000)
B. Household-head characteristics					
Age		−0.003	0.000	0.002	
		(0.002)	(0.002)	(0.003)	
Male		−0.043	−0.193	−0.073	
		(0.104)	(0.103)	(0.122)	
Married		0.113	0.133	0.053	
		(0.099)	(0.097)	(0.131)	
Educational attainment		0.003	0.014*	0.026	
		(0.006)	(0.006)	(0.015)	
Worked in the agricultural sector		−0.319**	−0.249**	−0.287*	
		(0.081)	(0.081)	(0.132)	
Worked in the non-agricultural sector		−0.145	−0.152	−0.222*	
		(0.087)	(0.084)	(0.094)	
Household size		0.030	0.035*	0.046	
		(0.017)	(0.016)	(0.028)	
C. Village characteristics					
Rural area			−0.169*		
			(0.069)		
Population (thousand)			0.043**		
			(0.012)		
Size (thousand hectares)			−0.084		
			(0.046)		

(*continued*)

Table 5.3 (continued)

Dependent variable: How much would you like to receive per hectare as rent if you like to relocate but do not want to sell your land?

	(1)	(2)	(3)	(4)	(5)
Size of non-agricultural land (thousand hectares)			0.728		
			(0.383)		
Distance to the nearest bus terminal (kms)			0.015**		
			(0.004)		
Time it takes to the nearest bus terminal (hours)			−0.164		
			(0.127)		
Distance to the district capital (kms)			−0.009**		
			(0.003)		
The time it takes to the district capital (hours)			0.794**		
			(0.139)		
Asphalt or cement road			−0.563**		
			(0.081)		
Fixed effects					
Village fixed effects				Yes	
Household fixed effects					Yes
Adjusted R-squared	0.049	0.061	0.146	0.055	0.045
The number of observations	1882	1877	1877	1877	1882

Note: The figures in parentheses are robust standard errors. * and ** indicate statistical significance at levels of 5 percent and 1 percent, respectively

to offer compensation which, from the landowners' perspective, is more reasonable as it accounts for the landowners' perspective.

In moving infrastructure projects forward, the government needs to give attention to creating a sense of fairness among the parties involved in the transfer of land. This also corresponds directly to the spirit of the Land Acquisition Law, which states that acquisition of land for public interest shall consider the balance between development interests and the interest of the public (Law No. 2/2012: Article 19).

A sufficient and fair amount of compensation for each individual is not the only means of assuring the well-being of affected people. It is also important to ensure that the community is well off. As suggested in "Voluntary Guidelines on the Responsible Governance of Tenure" (Food Agriculture Organization 2012), the rehabilitation of local roads and irrigation systems is crucial. Also, in many cases, the remaining farm plots are fragmented after the land transfer. Therefore, it is recommended that the central or local government take the initiative at consolidating the remaining arable land so long as it does not harm the risk reduction or crop diversification and if it does allow farmers to continue or increase productions. At the same time, it is indispensable to have better communication with community members and cultivate awareness of the benefits public infrastructure projects bring. It is crucial to share the vision behind a project and encourage participation of local communities so that no one will be left behind.

It is our hope that our findings can contribute toward designing a land acquisition scheme which balances development objectives, property rights, and social well-being, in order to support the development of infrastructure and, ultimately, the well-being of Indonesian citizens.

Notes

1. Such as Singapore 5th, Malaysia 32nd, and Thailand 45th (The World Bank 2016).
2. Referred English translation for Iqubal Rafani (2015) Strategic Plan of Indonesian Ministry of Agriculture: 2015–2019.
3. Under the Yudoyono administration, the 8.2 percent of the national budget is assigned for infrastructure (7.63 trillion Rp) in 2009. And it increased to 8.7 percent (17.8 trillion Rp) in 2014, 14.2 percent (2.9 trillion Rp) in 2015, 15.2 percent (3.17 trillion Rp) in 2016, and 18.6 percent (38.7 trillion Rp) in 2017.
4. 2017 April, fourth in Hotel *Kempinski*, Jakarta, when the Indonesian government launched a new land acquisition scheme for national strategic projects. The other issues are protracted planning and procurement (25 percent), limited funding (17 percent), and licensing (12 percent).
5. The presidential power of the expropriation of people's land rights that his 2005 decree and Law 20/1961 permitted as a democratically elected president (Davidson 2016).
6. The government had the legal domain over the nullification of land rights, local government officials were in charge of enforcing the state's policy on the ground under the 1999's decentralization law.
7. However, the agency was not clearly named as the National Land Agency (BPN).
8. Presidential decree: No. 71 (2012), No. 40 (2014), No. 99 (2014), and No. 30 (2015).
9. "Peraturan Kepala Badan Pertanahan Nasional Republik Indonesia Nomor 5 Tahun 2012 Tentang Petunjuk Teknis Pelaksanaan Tanah", October 2012 National (Decree of the head of the National Land Agency of the Republic of Indonesia, number 5, year 2012, Technical Guidelines for Implementation of Land Procurement).

10. For example, on acceleration of priority infrastructure provision (No. 75, 2014), on public private partnership projects (No. 38, 2015).
11. "Indonesia Valuation Standard" (*Standar Penilaian Indonesia/* SPI 306) *Penilaian Terhadap Pengadaan Tanah Bagi Pembangunan Untuk Kepentingan Umum.*
12. Wetlands are usually located lower than 200 meters below sea level, mostly in coastal areas, are downstream, and are passed by at least one river. The farm land in wetland areas is usually cultivated with paddy. Non-wetland areas are present more than 200 meters above sea level, are mountainous, and hilly areas. They are located upstream or near waterfalls. Farms in non-wetland areas are in the form of plantations.

References

ADB. (2017). *Report on the equivalence and acceptability assessments for the State Electricity Company of Indonesia (PLN): Consultation draft.*

Aziz, M. F. (2017). Indonesia should make land acquisitions more transparent and participatory. The Conversation, December 16, 2016, updated December 6, 2017.

Banerji, S., & Ghatak, M. (2009, September 30). No way out of this plot. *Financial Express.*

Capozza, D. R., & Helsley, R. W. (1989). The fundamentals of land prices and urban growth. *Journal of Urban Economics, 26*, 295–306.

Capozza, D. R., & Li, Y. (2002). Optimal land development decisions. *Journal of Urban Economics, 51*, 123–142.

Capozza, D. R., & Sick, G. A. (1994). The risk structure of land markets. *Journal of Urban Economics, 35*, 297–319.

Chicoine, D. L. (1981). Farmland values at the urban fringe: An analysis of sales prices. *Land Economics, 57*, 353–362.

Davidson, J. S. (2016). Eminent domain and infrastructure under the Yudhoyono and Widodo administrations. In J. F. McCarthy & K. Robinson (Eds.), *Land and development in Indonesia: Searching for the people's sovereignty* (pp. 167–185). Singapore: ISEAS-Yusof Ishak Institute.

Elad, R. L., Clifton, I. D., & Epperson, J. E. (1994). Hedonic estimation applied to the farmland market in Georgia. *Journal of Agricultural and Applied Economics, 26*, 351–366.

Fang, Y., Shi, K., & Niu, C. (2016). A comparison of the means and ends of rural construction land consolidation: Case studies of villagers' attitudes and behaviour s in Changchun City, Jilin Province, China. *Journal of Rural Studies, 47*, 459–473.

Food and Agriculture Organization. (2012). *Voluntary guidelines on the responsible governance of tenure of land fisheries and forests in the context of national food security*. Food and Agriculture Organization of the United Nations.

Ghatak, M., & Ghosh, P. (2011, October 8). The land acquisition bill: A critique and a proposal. *Economic & Political Weekly, XLVI*(41), 65–72.

Hardie, I. W., Narayan, T. A., & Gardner, B. L. (2001). The joint influence of agricultural and nonfarm factors on real estate values: An application to the Mid-Atlantic region. *American Journal of Agricultural Economics, 83*, 120–132.

Heming, L., & Rees, P. (2000). Population displacement in the Three Gorges reservoir area of the Yangtze River, central China: Relocation policies and migrant views. *International Journal of Population Geography, 6*(6), 439–462.

Hushak, L. J., & Sadr, K. (1979). A spatial model of land market behavior. *American Journal of Agricultural Economics, 61*, 697–701.

Jakarta Post. (2011, April 27). Editorial: Land control. *Jakarta Post.*

Masterson, T. (2007). Land rental and sales markets in Paraguay. *The Levy Economics Institute Working Paper No. 491*. Retrieved from https://papers.ssrn.com/sol3/papers.cfm?abstract_id=964906

Mendelsohn, R., Nordhaus, W. D., & Shaw, D. (1994). The impact of global warming on agriculture: A Ricardian analysis. *American Economic Review, 84*, 753–771.

Mills, D. E. (1981). Growth, speculation, and sprawl in a monocentric city. *Journal of Urban Economics, 10*, 201–226.

Ministry of Trade Republic of Indonesia, Kementerian Perdagangan Republik Indonesia. (2016). Kunerja Logistik Indonesia 2016.

Muryanto, B. (2018, April 11). PLN maintains blackout in Kulon Progo. *The Jakarta Post.* Retrieved from https://www.thejakartapost.com/news/2018/04/11/pln-maintains-blackout-in-kulon-progo.html

O'Rourke, K., & Milne, P. (2010). *Who's who in Yudhoyono's Indonesia: A socio-legal study on land, decentralization, and the rule of law in Bandung*. Leiden: Leiden University Press.

Plantinga, A. J., & Miller, D. J. (2001). Agricultural land values and the value of rights to future land development. *Land Economics, 77*, 56–67.

Quigg, L. (1993). Empirical testing of real option-pricing models. *Journal of Finance, 48*, 621–640.

Rafani, I. (2015). *Strategic plan of Indonesian Ministry of Agriculture: 2015–2019.* Indonesian Center for Agriculture Socio Economic and Policy Studies (ICASEPS), Ministry of Agriculture, Indonesia.

Shi, Y. J., Phipps, T. T., & Colyer, D. (1997). Agricultural land values under urbanizing influences. *Land Economics, 73*, 90–100.

Shonkwiler, J. S., & Reynolds, J. E. (1986). A note on the use of hedonic price models in the analysis of land prices at the urban fringe. *Land Economics, 62*, 58–61.

Tegene, A., Wiebe, K., & Kuhn, B. (1999). Irreversible investment under uncertainty: Conservation easements and the option to develop agricultural land. *Journal of Agricultural Economics, 50*, 203–219.

Titman, S. (1985). Urban land prices under uncertainty. *American Economic Review, 75*, 505–514.

Turvey, C. (2002). *Can hysteresis and real options explain the farmland valuation puzzle?* Department of Agricultural Economics and Business, University of Guelph.

Vitaliano, D. F., & Hill, C. (1994). Agricultural districts and farmland prices. *Journal of Real Estate Finance and Economics, 8*, 213–223.

World Bank. (2007). *Compiled reports of World Bank land acquisition research in Indonesia*. Washington, DC: World Bank. Retrieved from http://documents.worldbank.org/curated/en/569411468257988257/Compiled-reports-of-World-Bank-land-acquisition-research-in-Indonesia

World Bank. (2016). *Connecting to compete 2016: Trade logistics in the global economy—The Logistics Performance Index and its indicators*. Washington, DC: World Bank.

World Economic Forum. (2016). *The Global Competitiveness report 2016–2017: Insight report*. World Economic Forum.

CHAPTER 6

The Political Economy of Caste, Forced Displacement and Impact on Local Communities' Welfare in Nepal

Lai Ming Lam, Basant Pant, and Vengadeshvaran Sarma

Caste, Forced Displacement and Changes in Landownership

In this chapter, we explore the interplay between caste and forced displacement with a particular focus on the changes occurring in landownership and the long-term impact on the welfare of displaced communities. Caste is believed to be one of the most influential factors structuring the economic, political and social relations among various social groups in South Asian societies. Dumont (1980) argues that the concept of "caste" is associated with the practices of ranking the Hindu people on the basis of their perceived purity and impurity. In this sense, caste constructs visible

L. M. Lam
Osaka University, Suita, Japan

B. Pant
International Centre for Integrated Mountain Development (ICIMOD),
Patan, Nepal

V. Sarma (✉)
University of Nottingham, Semenyih, Malaysia

N. Yoshino, S. Paul (eds.), *Land Acquisition in Asia*,
https://doi.org/10.1007/978-981-13-6455-6_6

social rankings. Scholars such as Beall and Piron (2005, p. 9) and Gaige (1975) point out that caste "prevents individuals or groups from full participation in social, economic and political life and from asserting their rights". Nepal is no exception; the country's caste system was codified by the Legal Code (Muluki Ain) of 1854. The Code categorised the population into five caste groups with different social statuses. The twice-born Hindu castes were accorded the highest status. The *Newars* in the Kathmandu Valley and a few hill tribes shared the second highest caste ranking with representatives of the state. The remaining ethnic groups from the hills and Tarai were classified as enslavable alcohol drinkers (*masinya matawali*). Although the Code was officially abandoned in 1963 and some scholars also argue that this religion-inspired categorisation is too ideological and may conflict with local social rank (Quigley 1993, p. 102), caste still plays an important role throughout Nepalese society; our current study site is no exception (Lam 2014).

Since the late 1980s, human displacement as a conservation strategy has been commonly practised around the world, particularly in developing countries, despite the increasing criticism by social scientists that this represents a severe human rights violation. Social scientists argue that protected areas have most often been created in regions where the most socially, economically and politically vulnerable groups reside (Agrawal and Redford 2009). In other words, conservation-induced displacement may seriously damage the traditional yet powerless social groups. Scholars point out that fundamental features of displacement involve not just loss of place and material assets but also losses of social ties, identity and meaning (Scudder 2005; De Wet 2006; Downing and Garcia-Downing 2009).

In the past half century, a new sociopolitical and economic landscape has emerged in the frontier lowland Tarai region in Nepal. Due to the success of the malaria eradication programme, Tarai is no longer the homeland of the indigenous *Tharu* groups but is now inhabited by different caste/ethnic groups from the hill areas (called *Pahadia*). Recent studies on different *Tharu* groups have shown that land is a contentious issue in contemporary *Tharu* society. For example, McDonaugh (1997, 1999) and Odegaard (1997) have shown that control over land has shifted significantly from the hands of the *Dangauras* to the migrant *Pahadias*; this has impacted *Dangaura* people's livelihoods adversely. Many *Dangauras* have firstly become *kamaiya* (bonded labourers), and further, many of them have been moving to far west Tarai to escape over-exploitation by the *Pahadia* landlords. A similar observation is made by new relationships between them and the *Chitwan Tharus* as the latter lost their land. Apart from the transfer of landownership, *Tharus* have also faced the livelihood challenges caused by the establishment of protected areas in the Tarai region. Forced displacement was carried out in Chitwan National Park, Bardiya National Park and our current study site Shuklaphanta National Park (then Shuklaphanta Wildlife Reserve; hereafter referred as the park) where the largest resettlement programme was implemented. So how does the forced displacement together with ongoing migration affect the livelihoods of different ethnic/caste groups living adjacent to the park? Are these impacts heterogeneous across ethnic and caste lines? Especially, is the indigenous *Tharu* group suffering the most? How do the displaced communities cope with the livelihood challenges in the post-displacement period? Does forced displacement have a long-term impact on the welfare of those displaced communities?

In this chapter, we continue to use the interdisciplinary approach—"Anthronomics"—which relies on anthropological fieldwork and large quantitative surveys to better understand the livelihood changes in the post-displacement period. We successfully interviewed 830 households from different ethnic and caste backgrounds living adjacent to the park, either affected or not affected by the displacement. The diversified sample enables us to track the heterogeneity in coping strategies so that risks could be overcome. Our analysis focuses particularly on the landownership status in this region and the coping mechanisms people employ in association with forced displacement due to the park extension programme.

A Description of Local Communities

There are four major social groups residing in the region of our study. *Rana Tharus* (hereafter referred to as *Ranas*) first settled in Rauteli Bichawa village which is located in the central part of the current park and was believed to be the first human settlement in Kanchanpur district. Although stories vary on the origins of the *Ranas* (see Lam 2011), it was clear that they came from India and settled in Rauteli Bichawa village few generations long before the establishment of the park. They relied on agriculture, and most of them were big landlords (*Jimidars*) in the region and village leaders.

Dangaura Tharus (hereafter referred to as *Dangauras*), who also called themselves Chaudhary, are people who started moving out from western Tarai to the park area in 1950 because of severe exploitation from their new *Pahadi* landlords. Their major livelihood was agriculture, but they owned less land compared to the *Ranas*, as the former were latecomers. Although *Dangauras* and *Ranas* were categorised by the state as constituting the same social group (*Tharus*), their interactions were rare. For instance, they did not practise intermarriage and continually perceived themselves to be very different from each other.

Pahadia is the term referring to people who originated from the hill and mountain areas, and they comprise different caste/ethnic groups. The highest (social rank) and largest caste groups are the Brahmin and *Chhetri*. According to our *Rana* informants from Rauteli Bichawa, any interactions with *Pahadias* were occasional but positive before the hill migration. At that time, *Pahadias* perceived the Tarai area as being dangerous due to the widespread occurrence of malaria; therefore, they only visited Kanchanpur in the dry seasons for casual labouring work and trading food products with the *Rana* community. However, the situation changed dramatically in the 1950s when the state's intervention in land policy became more obvious and *Pahadias* also started to migrate to Kanchanpur.

Landownership Transformation in the Park Region

Since the 1950s, the state wanted to reassert its authority over the country's natural resources at all levels of government. One of the major policies to make this possible was to centralise state control over local land resources through new land reform programmes (Pfaff-Czarnecka 1997, p. 437; Sharma 1997, p. 479). Our local informants pointed out that since the 1960s, in Rauteli Bichawa village, the local *Ranas* started to deal directly with the state instead of the *Rana* landlords regarding land issues. Based on the life histories collected from the Rauteli Bichawa villagers, many government officials came to their village to map their land and establish new settlements. Consequently, officials visited Rauteli Bichawa village throughout the 1970s and 1980s, and one of the changes resulting from state intervention was to diminish local autonomy by removing the right of the *Ranas* as landlords. If *Ranas* did not register land following the new procedure, they did not own any land. The *Ranas*, like many traditional societies, failed to perceive that landownership in modern times was an exercise represented by possession of land registration documents. For them, the concept of landownership was entangled with cultivating land for generations, so they never feared losing it. This, however, put the indigenous *Rana* population into a weaker position of not being able to protect their ancestral land through registration, particularly those who had small landholdings.

Meanwhile the state played a leading role in the transformation of landownership from the hands of indigenous *Ranas* to the *Pahadias*. This was administered through a series of land reform policies and state-sponsored resettlement programmes in Kanchanpur, particularly in the Rauteli Bichawa village and in the Terai region as a whole. The migrants were mainly higher caste families, specifically Brahmins and *Chhetris*, who were mostly literate and had closer ties with the equally literate state officials due to the same higher level of literacy and culture. This made access to land resources and assistance from the state easier for them to acquire and in turn gave them greater control over the land. Our previous fieldwork has shown that the Kanchanpur district had experienced dramatic changes in demography and landownership over the past few decades. Bhattarai, Pant and Timalsina's latest study (2017) also documents that continuing population growth due to the hill migration has put pressure on scarce land resources in the park area.

Lam et al. (2016) indicate that the *Ranas* received the largest land compensation compared to other ethnic/caste groups, the study was mainly carried out in Dhokka Block where previous big landlords dominated the resettlement site. Such micro-level data limit our comprehension of the larger picture concerning the impact of displacement on people's welfare, coincidentally with ongoing demographic and landownership changes surrounding the park area.

Research Methods

To address our research questions more effectively, we not only went back to the same villages but also expanded our analysis to other non-displaced villages around the park. Our primary quantitative data come from questionnaires administered to 830 households. We interviewed 509 displaced households, which is about 61% of the full sample (our treatment group). The rest (321 households) are non-resettled (our control group) and come from Rauteli Bichawa (179), Pipladi (92) and Dekhatbhuli (50). The displaced households are predominantly from Rampur (198) and Jhalari (148), followed by Kalika (83), Laxmipur (71) and Beldandi (9). Overall, the distribution is justified as we aim to evaluate the long-term impact of displacement on affected households' coping strategies. All our data are collected at the household level since our research goal is to identify "household" coping mechanisms. Additionally, we also conducted focus group discussions; this comprises key informant interviews and observations in the studied villages to collect qualitative information that answer the overarching research questions. A diversified sample would enable us track the heterogeneity in coping strategies to mitigate risks. Linking these outcomes to the sense of social exclusion, which was historically rooted in the Nepalese government's land settlement policies (Lam 2011), would enrich our understanding of how development can proceed in its myriad forms.

Given the ethnic/caste-centric social life that still prevails in Nepal, we regrouped different caste/ethnic communities into four broad categories: *Rana* (306), Brahmin, *Chhetri* and *Newar* (BCN) (336), *Dangaura* (158) and Dalit (30). The distribution of displaced and non-displaced households across these broad ethnic/caste groups is homogeneous. This implies that there is no ethnic group-specific bias in the sample (see Tables 6.1 and 6.2).

Table 6.1 Study site locations

	Study villages	*No. of interviewed households*	*Total*
Resettled village	Rampur	198	509
	Belandi	9	
	Jhalari	148	
	Laxmipur	71	
	Kalika	83	
Non-resettled village	Rauteli Bichawa	179	321
	Piladi	92	
	Dekhatbuli	50	
	Total	*830*	*830*

Table 6.2 Resettled/non-resettled households by major ethnic/caste groups

Ethnic/caste group	*Non-resettled*	*Resettled*	*Total*
BCNs	44.86	37.72	40.48
Ranas	30.22	41.65	37.23
Dangauras	22.74	16.70	19.04
Dalits	2.18	3.93	3.25
Total	*100.00*	*100.00*	*100.00*

Results

We constructed the adult-equivalent per capita household consumption expenditures for each household by aggregating expenditures on the broad consumption categories, a standard indicator of well-being employed in economic analyses. We compared the average value of this welfare indicator across the ethnic/caste groups. Surprisingly, Dalits and BCNs indicate almost the same average level of expenditure (about 27,500 Nepalese Rupees a year) followed by the *Ranas* (21,600) and the *Dangauras* (19,000) (Table 6.3). It could be because we interviewed a biased sample of only rich Dalit households, since there are only 27 displaced Dalit households. Once we make comparisons within each caste group, displaced households on average report a lower welfare level compared to non-resettled households from all ethnic groups, except the *Dangauras*. For the *Ranas* and Dalits, the average welfare indicator is significantly lower for displaced households compared to non-displaced households; however, for the BCN group, the difference in the average welfare level between the displaced and non-displaced households is negligible. The results may have two implications: firstly, the displacement did not wield a significant impact on BCNs or, secondly, BCNs developed better coping mechanisms to changes in their livelihoods after displacement.

As stated previously, whoever has control over land resources can play the key or driving role in local economic and political power relationships. In Nepal, land was the property of the state, and this type of land was known as *Raikar*. Under this state-as-landlord system, the government had absolute power to grant and confiscate land for grantees and could retain appropriate land for its own needs (Caplan 1970). All *Raikar* users had only the right to use land but not the right to alienate any part of it, or to sell it or mortgage it. One of the major changes created by the park resettlement compensation scheme was that the state could grant the land compensation and title for eligible households. According to our household survey, the average percentage of household landholding with a title

Table 6.3 Average per capita adult-equivalent expenditure by caste and resettlement status (in Nepali Rupees)

	BCNs	Ranas	Dangauras	*Dalits*
Non-resettled	27,993.75	26,393.89	17,067.82	32,157.88
Resettled	27,081.81	19,406.88	20,810.28	25,849.13

is significantly higher for the displaced households (about 88%) compared to the non-resettled households (approximately 49%). This implies that upon displacement, households became more empowered in terms of having a title to their land. Our data show that the percentage of households with land titles was significantly higher among resettled households across all caste groups. This particularly affected resettled Dalit households, who recorded an 89% higher land title possession record compared to 14% among the Dalit households that were not affected (Table 6.4).

Although land title is given to displaced households from all ethnic/caste groups, our interviews and field observations support the contention that some socially powerless groups found it difficult to collect land registration documents. Therefore, they did not receive any land compensation or land title. For example, many *Rana* informants pointed out that they did not receive full land compensation as some of their previous land was "non-registered". Some *Ranas* were also denoted as landless (when in reality they were not) and only received two to ten *kattas* of land. As a result, displaced *Rana* households had significantly smaller landholdings compared to non-displaced *Rana* households (Table 6.5).

Table 6.5 also shows that landholdings among affected households were about six *kattas* smaller than non-affected households. It is evident from the summary in Table 6.6 that even with smaller landholdings owned by displaced villagers, the yields from the compensated land were significantly lower, with the exception of the displaced Dalits. All other major caste groups reported a decrease in yield. Displaced *Dangauras* specifically reported a 58% smaller yield compared to non-displaced *Dangauras*.

Table 6.4 Percentage of households with land titles by caste

	BCNs	Ranas	Dangauras	*Dalits*
Non-resettled	0.25	0.73	0.68	0.14
Resettled	0.90	0.89	0.79	0.89

Table 6.5 Average current landholding size (in *katta*) by caste

	BCNs	Ranas	Dangauras	*Dalits*
Non-resettled	15.6	31.7	21.9	8.3
Resettled	14.3	24.6	14.8	9.5

The resultant loss of land and low yield meant that the displaced households experienced less food security as evident in Table 6.7. Table 6.7 reports the proportion of households that reported having sufficient food for at least six months.

From Table 6.7, it can be seen that, on average, resettled households reported 20 percentage points lower food availability (for a period of at least six months) compared to non-resettled households. The disparity was largest among *Dangauras*. Only 3.5% of the resettled *Dangauras* reported having food for at least six months, compared to 52% of the non-resettled *Dangauras*. The only exception was Dalits; resettled Dalits had better food sufficiency compared to non-resettled Dalits; however, this result has to be interpreted with caution due to the lower sample size of Dalits.

The statistics reported in Table 6.8 explain the possibly significant increase in the number of means to make a living among affected households.

Table 6.6 Low yield by caste

	BCNs	Ranas	Dangauras	*Dalits*
Non-resettled	0.60	0.49	0.15	0.86
Resettled	0.67	0.51	0.73	0.35

Table 6.7 Food security index by caste (% of households reporting food sufficiency for six months or more)

	BCNs	Ranas	Dangauras	*Dalits*
Non-resettled	32.64	41.24	52.05	0.0
Resettled	21.35	22.17	3.53	20.0

Table 6.8 Livelihood sources over time by caste

		BCNs	Ranas	Dangauras	*Dalits*
2001	Non-resettled	2.39	2.61	2.92	2.43
	Resettled	1.99	2.64	2.75	2.45
2008	Non-resettled	2.62	2.77	2.97	2.43
	Resettled	2.30	2.87	2.94	2.75
2013	Non-resettled	2.77	2.99	2.99	2.43
	Resettled	2.43	3.06	3.06	3.30

Table 6.9 Provision of agricultural labour on other farms over time by caste

		BCNs	Ranas	Dangauras	*Dalits*
2001	Non-resettled	0.48	0.89	0.77	0.71
	Resettled	0.55	0.64	0.84	0.50
2008	Non-resettled	0.46	0.68	0.73	0.71
	Resettled	0.58	0.69	0.88	0.50
2013	Non-resettled	0.47	0.69	0.74	0.71
	Resettled	0.60	0.70	0.88	0.58

The number of livelihood sources rose from an average of 2.4 during pre-resettlement to 2.8 in 2013 for the affected households. This was comparable to an average of 2.7 sources during pre-settlement to 2.9 sources in 2013 for unaffected households. Dalits (socially the lowest caste group) increased the number of their livelihood sources by almost one, which suggests that each Dalit household had on average one additional working member/job over a ten-year span.

Summary statistics also reveal that agricultural labour provided for other people's farms increased by 5% across all resettled households, whilst the percentage remained almost unchanged across the unaffected households. The exception to this was the *Ranas*, whose proportion of labour provided in other farms fell by a staggering 20%. Therefore, it could be read that displaced households worked as agricultural labourers in other people's farms as a coping mechanism to overcome food insecurity (Table 6.9).

Our data also show that while access to electricity significantly increased over the past decade across households, access to electricity was on average 20% lower amongst affected households compared to unaffected households; the worst affected were *Rana* households with an enormous 40% difference. Similarly, data also reveal a significant improvement in access to health care facilities and public transport across households; however, the inter-caste differences were remarkably high. One possible explanation for these differences is the clustering of caste members in particular villages, and a second possible reason is the socio-economic status of households being strongly correlated to their caste status.

Results from our basic Ordinary Least Squares (OLS) output (reported in Table 6.10) show that in the long run, the effect of displacement and caste dynamics do not play a significant role in determining the per capita expenditure (a proxy used to measure welfare of households). Our results

Table 6.10 OLS results for the effect of caste and resettlement status on households' welfare

Dependent variable	*Log per capita expenditure*
Ranas	0.0037 (0.1354)
Dangauras	−0.0537 (0.1248)
Dalits	0.0831 (0.1366)
Resettled	−0.0542 (0.1719)
Resettled × *Ranas*	−0.0427 (0.1573)
Resettled × *Dangauras*	0.0226 (0.1415)
Resettled × Dalits	−0.2238 (0.1535)
Other controls	✓
Observations	721
R^2	0.4534

Note: The reference ethnic/caste group is BCN. The other controls include gender, age and education level of the head of the household, number of children, asset holdings, access to facilities and harvest yield information. Robust standard errors are reported in parentheses.

show that even the interaction terms of displacement and caste are not statistically significant, an indication that the effect of displacement is rather homogeneous across the castes.

Discussion

For the large-scale household survey, we did not find a significant difference in livelihood and welfare based on ethnicity or caste. However, the result indicates ongoing changes in landownership and livelihood coping strategies among different ethnic/caste groups. First, the *Ranas* were historically supposed to be the biggest landlord group among others, and in fact they were; however, the difference between resettled and non-resettled *Ranas* was large. A similar situation was true of another *Tharu* group, the *Dangauras*. Second, displacement is expected to cause both physical and psychological losses to affected households, which is evident by the welfare indicators between displaced and non-displaced houses shown earlier. The more severe outcomes were found for the indigenous *Ranas* and *Dangauras* (Table 6.3). Surprisingly, there was no significant difference between displaced and non-displaced BCN groups. This indicates that the impact level of displacement on displaced communities varied by ethnicity.

These findings imply that displacement also acts as an important trigger that reproduces existing inequality. As discussed earlier, although indigenous inhabitants were granted the right to get full land compensation, in reality, their weak political and cultural capital made the *Tharu* groups (i.e. *Ranas* and *Dangauras*) undergo enormous pressures on their claim to land rights. For example, previous big *Rana* landlords complained that they lost their unregistered ancestral lands; the small landholders were often categorised as landless and lost all of their lands following displacement. Some *Ranas* pointed out that the government often ignored their grievances. Many *Ranas* complained that in many instances, the disputes over land between them and the migrants were resolved in favour of the latter. In the matter of registration of and transaction concerning land, good verbal and written communication skills were required when conducting business with government officials. There were also complaints against the migrants for taking advantage of the illiterate *Ranas* and confiscating their lands by providing them with flawed contracts. For example, one displaced *Rana* stated that without the informed consent and authorisation of his grandfather, his father signed a land transaction document to a migrant state official. However, when his grandfather contested it, the land had already become a property of the state official. It explains why the displaced *Rana* and *Dangaura* households own less land, on average, than non-displaced households.

Apart from landownership, we can note that the BCN group has the highest flexibility and adaptability in terms of livelihood diversification. Our data clearly show that displaced households have developed more means of obtaining a livelihood after the displacement, and here the BCN and Dalit groups were the most outstanding groups. With land compensation and new livelihood strategies, the consumption expenditures between displaced BCN and non-displaced BCN households exhibited no discernible difference. The finding implies that displaced BCN households were much more resilient than other social groups, and they did not lose substantial amounts of land caused by displacement. Yet, like all the other social groups, they developed more ways to diversify their livelihoods. As a result, displaced BCN households retained similar livelihood standards as those that were not affected by displacement.

Although the above discussion suggests that there is no single explanation for how displacement affects local communities' material welfare in the long term, it is evident that the trajectory of local history, socioeconomic structure and the capacity of displaced households, all influence the outcomes. Furthermore, it is clear that the social and psychological impacts of displacement on communities are long term and very difficult

Table 6.11 Emotional well-being by resettled and non-resettled households by caste

		Happy now? (1—Very unhappy, 5—Very happy)		*Income ladder (1—Very poor, 5—Very rich)*	
		2001	*2013*	*2001*	*2013*
Non-resettled	*Ranas*	2.73	2.69	2.50	2.54
	Dangauras	2.14	2.10	2.10	2.16
	BCNs	2.11	2.30	1.70	2.60
	Dalits	1.86	1.86	1.86	1.86
Resettled	*Ranas*	1.59	1.58	1.55	1.57
	Dangauras	1.15	1.49	1.15	1.43
	BCNs	1.38	1.46	1.38	1.54
	Dalits	1.67	1.67	1.67	1.67

to overturn. Our current study echoes previous analyses that displacement has severely disrupted the social life of affected communities (Scudder 2005; De Wet 2006; Downing and Garcia-Downing 2009). The self-reported emotional well-being indicator shows that displaced households from all caste/ethnic groups felt unhappier and poorer than non-displaced households (Table 6.11). Negative feelings about their livelihood have not been improved in the ten years following displacement. This clearly indicates that material welfare did not have a significant impact on people's emotional well-being. The experience reminds us that displacement should not be taken up as a viable option unless it is the last resort. If displacement is inevitable, the "cure" for this should not simply be a materialistic-based compensation scheme. Instead, policymakers must provide psychological and counselling services to the affected communities.

Conclusion

In this chapter, using both quantitative and qualitative data, we show that the impact of displacement on communities' welfare is multidimensional and dynamic, leading to interactions with embedded socio-economic and political factors. Displacement is an impactful process that reshapes the local socio-economic system and conservation-induced displacement, in particular, and tends to exert a negative impact on traditional powerless social groups.

Although the household survey data do not provide sufficient support to the assertion that displacement affected communities' welfare differently according to ethnicity, the changes in control over land resources before and after displacement gradually emerging in the park region have continuously reshaped the local economic, political and social relationships and dynamics. This study shows that the indigenous *Ranas'* landlord status is being challenged due to the displacement process. Due to less political clout and lower social status, Ranas did not receive full land compensation as promised in the Nepalese government's resettlement policy and some were even categorised as landless. The loss of economic power implies the further political and social marginalisation of indigenous groups like the Ranas and *Dangauras*. Consequently, displacement has widened the social and economic gap between the indigenous *Ranas* and the dominant BCN group.

In the case of the park conservation exercise, it is obvious that the land compensation scheme did not restore people's livelihoods and/or incomes because the enacted policies failed to take into account social divisions, particularly the control and ownership of land resources, which have long been embedded in Nepalese society. Indigenous social groups are often the victims, since compensation policies only take into account present-day entitlements without considering the history and cultural transformations over time. This also indicates the necessity for a land compensation framework that accounts for the social divisions and political economy of past land settlement policies. Without thoughtfully considering the political, economic and cultural contexts, compensation schemes (especially financial compensation schemes) will only serve as a mechanism that further accelerates social inequality and social strife among different groups.

References

Agrawal, A., & Redford, K. (2009). Conservation and displacement: An overview. *Conservation and Society, 7*(1), 1–10.

Beall, J., & Piron, L. (2005). *DFID social exclusion review*. London: London School of Economics and Political Science.

Bhattarai, S., Pant, B., & Timalsina, N. (2017). Conservation without participation: Detrimental effect of escaping people's participation in protected area management in Nepal. In S. A. Mukul & A. Z. M. M. Rashid (Eds.), *Protected areas: Policies, management and future directions* (Chap. 4, pp. 83–104). New York: Nova Science Pub. Inc.

Caplan, L. (1970). *Land and social change in east Nepal: A study of Hindu-tribal relations.* London: Routledge & Kegan Paul.

De Wet, C. (2006). Risk, complexity and local initiative in forced resettlement outcomes. In C. De Wet (Ed.), *Development-induced displacement: Problems, policies and people.* New York: Berghahn Books.

Downing, T. E., & Garcia-Downing, C. (2009). Routine and dissonant cultures: A theory about the psycho-socio-cultural disruptions of involuntary displacement and ways to mitigate them without inflicting even more damage. In A. Oliver-Smith (Ed.), *Development and dispossession: The crisis of forced displacement and resettlement.* Santa Fe, NM: School for Advanced Research Press.

Dumont, L. (1980). *Homo hierarchicus: The caste system and its implications.* Chicago: University of Chicago Press.

Gaige, F. H. (1975). *Regionalism and national unity in Nepal.* Berkeley, CA: University of California Press.

Lam, L. M. (2011). Cultural perspectives of land and livelihoods: A case study of Royal Shuklaphanta Wildlife Reserve in the far-western part of Nepal. *Conservation and Society, 9*(4), 311–324.

Lam, L. M. (2014). Changes in Rana Tharu–Pahadia relationships: Land, politics and caste. In Y. Raj (Ed.), *Ruptures and repairs in South Asia: Historical perspectives* (Martin Chautari Book Series 78) (pp. 123–144). Kathmandu: Martin Chautari.

Lam, L. M., Paul, S., & Sarma, V. (2016). Reversal of fortune? The long-term effect of conservation-led displacement in Nepal. *Oxford Journal of Development Studies, 44*(4), 401–419. https://doi.org/10.1080/13600818.2016.1149158.

McDonaugh, C. (1997). Losing ground, gaining ground: Land and change in a Tharu community in Dang, West Nepal. In D. Gellner, J. Pfaff-Czarnecka, & J. Whelpton (Eds.), *Nationalism and ethnicity in a Hindu kingdom: The politics of cultures in contemporary Nepal* (pp. 275–298). Amsterdam: Harwood Academic Publishers.

McDonaugh, C. (1999). Aspect of social and cultural change in a Tharu village community in Dang, West Nepal, 1980–93. In H. O. Skar (Ed.), *Nepal: Tharu and Tarai neighbours* (pp. 223–234). Kathmandu: Educational Enterprises, Mandala Book Point, Ratna Pustak Bhandar.

Odegaard, S. E. (1997). *From castes to ethnic group? Modernisation and forms of social identification among the Tharus of the Nepalese Tarai.* PhD dissertation, University of Oslo.

Pfaff-Czarnecka, J. (1997). Vestiges and visions: Cultural change in the process of nation-building in Nepal. In D. N. Gellner, J. Pfaff-Czarnecka, & J. Whelpton (Eds.), *Nationalism and ethnicity in a Hindu kingdom: The politics of culture in contemporary Nepal.* Amsterdam: Harwood Academic Publishers.

Quigley, D. (1993). *The interpretation of caste.* New York: Oxford University Press.

Scudder, T. (2005). *The future of large dams: Dealing with social, environmental and political costs.* London: Earthscan.

Sharma, P. R. (1997). Nation-building, multi-ethnicity and the Hindu state. In D. N. Gellner, J. Pfaff-Czarnecka, & J. Whelpton (Eds.), *Nationalism and ethnicity in a Hindu kingdom: The politics of culture in contemporary Nepal.* Amsterdam: Harwood Academic Publishers.

CHAPTER 7

Special Economic Zones and Livelihood Changes: Evidence from India

Saumik Paul and Vengadeshvaran Sarma

Introduction

Special Economic Zones (SEZs) are typically designed to create jobs, promote exports and drive economic growth. However, in recent years, SEZs have stirred controversy and debate in India for different reasons. There has been resistance towards the setting up of new SEZs across India, particularly in the state of West Bengal, for being set up through forcible land grabs (Nielsen 2010). Recent studies estimate that over a million people will be displaced in India due to the setting up of SEZs in the coming years (Mansingh et al. 2012; Mukherjee et al. 2016). Moreover, only 1 in 18 will receive direct employment and an additional 1 in 13 will receive indirect employment, suggesting that most households affected by the setting up of SEZs in India are likely to experience loss of livelihoods. Academic and policy debates have thus centred on the trade-off between industrialisation-led mass welfare and marginal private cost to those

S. Paul (✉)
Asian Development Bank Institute (ADBI), Tokyo, Japan
e-mail: spaul@adbi.org

V. Sarma
University of Nottingham, Semenyih, Malaysia

N. Yoshino, S. Paul (eds.), *Land Acquisition in Asia*,
https://doi.org/10.1007/978-981-13-6455-6_7

directly disadvantaged by the setting up of SEZs. Two key issues that have received the most attention are the optimum compensation for the affected and the long-term effects of SEZs through changes in livelihoods. Banerjee et al. (2012) highlight the concerns regarding livelihood effects such as fulfilment of promised jobs to displaced households, and whether the benefits of industrialisation favour structural changes in livelihoods of the affected households, remain under-researched.

This chapter examines the livelihood changes of the affected households due to the establishment of the Falta Special Economic Zone (FSEZ, here on). The FSEZ was chosen as the subject of this study for three reasons. First, the FSEZ is located in West Bengal and was the first of its kind. Given the recent violent incidents of SEZ-led displacement followed by a sociopolitical upheaval in West Bengal, this study opts to focus more on the state. Second, the FSEZ has been a relatively successful SEZ in West Bengal (Shalti Research Group 2008). Most previous studies in the literature focus on failed SEZs on an ex post facto basis, which fail to sufficiently address the effects of SEZs on livelihoods. Third, being established in 1984, the FSEZ allows a sufficiently long period to assess its effects on livelihoods of those who are both affected and unaffected by its setup.

Based on existing empirical and anecdotal evidences, this study uses an analytical structure addressing the following research questions:

1. What is the likelihood of a displaced household member (1) being employed? And (2) being employed at the FSEZ?
2. Do members of displaced households enjoy similar returns to education and experience compared to other affected households? Does this result hold for those working at the FSEZ?
3. Does the FSEZ generate employment for women and contribute to reducing gender disparities in wage and education?

Since the overall effect of SEZs on livelihoods is likely to stretch over a number of years, it is imperative that the above questions are evaluated over a long period of time. Hence, the study focuses on the FSEZ. Structured questionnaires and focus group discussions (FGDs) were administered in three villages near the FSEZ. All directly affected households (630 in total, including displaced households and land acquired households) and a sample of 387 households (control group) that were indirectly affected by FSEZ were surveyed. Affected households comprise two groups: (1) displaced households—lost farmland and residences and had to move and resettle in a new place and (2) land acquired households—lost farmland but did not have to relocate. Land acquired households lost land that was used to accommodate the displaced households. Unaffected households endured indirect effects of the FSEZ. With the setting up of FSEZ, roads connecting to the city were developed and widened. This resulted in the loss of livelihood for those engaged in transporting goods via the river.

The empirical outcomes indicate long-term adverse livelihood effects for displaced households. The findings indicate lower labour market participation rates among affected household members. However, members from displaced households were 8% more likely to be employed within the FSEZ. Returns to education are positive among all three categories of households. However, returns to education are significantly lower for displaced households working within the FSEZ, which is predominantly driven by female employees. Among active labour market participants, more than half of the women from displaced and unaffected households worked in the FSEZ. The FGDs indicated the presence of an informal market for labour recruitment at the FSEZ, which hurt most of the villagers. Evidence also suggests that structural transformation from agriculture to manufacturing, induced by the setting up of the FSEZ, resulted in loss of traditional livelihoods for many villagers causing widespread structural unemployment.

Unobserved differences across the three categories of households, prior to the setting up of the FSEZ, can contribute to endogeneity, which invalidates the above results. To address potential endogeneity, the study uses the average education level of household members, 45 years and above (i.e. those who were at least 15 years old in 1984), to control for pre-FSEZ effects. This assumes that the education level of those who are currently 45 years and above was unaffected by the setting up of the FSEZ. This approach is aimed at minimising the problem of endogeneity in the regression framework. This approach is similar to unobserved effects or fixed effects model (Wooldridge 2009, p. 456). After controlling for pre-FSEZ characteristics, estimates indicate that wages are not statistically different across the three categories of households. Similarly, labour force participation rates and work participation rates at the FSEZ are no longer statistically significant. These estimates are robust to alternate specifications including clustering of standard errors at the household level, and inclusion of a proxy for income.

Overall, the long-term livelihood effects of the FSEZ on those (un) affected are mixed and highlight a quintessential equity-efficiency trade-off. On the one hand, displaced households are statistically equitable in their livelihood outcomes, compared to other groups of households. Over time, there is also evidence that access to human capital and returns to education show convergence across gender and household groups. On the other hand, all groups of households are likely to suffer from efficiency loss due to the structure of informal recruitment within the FSEZ and loss of traditional livelihoods. While these findings may not necessarily echo the livelihood outcomes of SEZs in other parts of the world, the policy implications arising from these findings may be broadly applicable to SEZs globally.

Background, Study Design and Data

In India, there are currently 205 SEZs in operation with a further 415 pending approval. The 205 SEZs employed over 1.5 million people in 2015 and contributed to about US$47 billion in exports. The first SEZ in India (then referred to as export processing zone) was set up in 1965. Until 2014, land acquisition for such SEZs was governed by the colonial era 1894 Land Acquisition Act. The act allowed the state to take over private land for public purposes (similar to eminent domain). One of the arguments in classifying SEZs as public purpose investments was their creation of jobs. While setting up SEZs in fallow land might lead to Pareto improvements, in almost all cases, SEZs were set up on arable land with better access to sea ports leading to so-called Pareto-efficient conditions (Kundra 2000). Such a Pareto-efficient setup, while potentially bringing benefits to the country as a whole, imposes private costs on agricultural farm owners, agricultural workers and households forced to relocate.

The FSEZ was set up in 1984 as the first SEZ of West Bengal and remains one of the few in the country to operate over three decades. The FSEZ was initiated with 72 firms—only 52 firms are currently operational. The FSEZ was established through acquisition of land from two major sources: the Calcutta Port Trust, a state enterprise, which contributed to more than 60% of the FSEZ land and the villages of Akalmegh and Uttar Simulbera, which contributed to about a third of the FSEZ land. Eighty acres of additional land was also acquired from a third village, Gopalpur, to resettle affected villagers from the aforementioned two villages of Akalmegh and Uttar Simulberia. While most villagers from Akalmegh and Uttar Simulberia (hereon referred to as displaced households) had to give up their residence and residential land, most villagers from Gopalpur (hereon referred to as land acquired households) had to give up their agricultural land for the settling of those displaced villagers (Roy 2008). We refer to both displaced households and land acquired households as directly affected households in later sections of the chapter.

Objectives and Criteria for Sampling

The objective for sampling was to have households in all the three categories: (A) displaced households, (B) land acquired households and (C) unaffected households. Figure 7.1 shows the geographical location of these categories of households. Households from category 1 are now resettled in the area identified as '2' and category B households are

Fig. 7.1 The villages studied in this chapter. Note: 1—Nainan, 2—New Gopalpur ('Higland') and 3—Gopalpur. Source: Google Maps and authors' own point reference

resident in the area identified as '3'. A complete census of these two villages—New Gopalpur (Highland) and Gopalpur, respectively, was conducted.

There are a number of villages within a 5-kilometre radius of the FSEZ that fit the criterion of category 'C'. However, given that the objective is to assess the effect of displacement on livelihood impacts, proximity to the FSEZ was vital. Distance to the FSEZ could affect employment prospects and the severity of effects by the setting up of FSEZ. Therefore, the sample was restricted to households within a 500-metre radius of the FSEZ. This narrows down to the northern half of the village 'Nainan' (identified as '1' in Fig. 7.1), which represents households in category 'C'. By measuring a 500-metre radius cut-off point, an imaginary border was drawn through the village of 'Nainan' for sampling purposes. The proximity of Nainan to Gopalpur and Highland indicates that Nainan shares similar characteristics to the affected villages in terms of infrastructure and access to facilities such as schools, health clinics, banks and local administrative offices.

Data

The data comprises 1017 households. This includes 462 households from the first category (those who were physically displaced), 168 from the second category (those whose land was acquired to resettle those in the first category) and 387 households that were not affected. At the individual level, there are 4780 observations.

Descriptive statistics (not presented here for brevity) show that individual and household characteristics are balanced (within 10% deviation from the mean) across the three categories of households, with the exception of education attainment. Members from land acquired households were older and more educated compared to those from displaced and unaffected households. Land acquired households had the highest per capita adult expenditure, followed by displaced households and by unaffected households. Other descriptive statistics relevant to the research questions are discussed in more detail later in the chapter.

Effects on Livelihood

Prior to the setting up of FSEZ, almost all households in the region worked in the primary sector—agriculture and fishing. Land acquired households engaged in agriculture and were agricultural land owners (48%); displaced households worked on some of these agricultural lands and engaged in fishing and boating but were not agricultural land owners (only 1% held agricultural land). Among unaffected households, 12% held agricultural land, and others mostly engaged in fishing and boating. While almost all of the land owners among land acquired and unaffected households held deeds for their agricultural land, only 25% of households among displaced households had a deed for their cultivable plot.

The establishment of the FSEZ caused livelihood changes across three channels: (1) loss of agricultural land (loss of agricultural livelihood for land owners and those who were employed on such land); (2) physical relocation (loss of access to traditional livelihoods such as boating and fishing); and (3) structural change (the emergence of the FSEZ required a change in skills from being employed in the primary sector to the manufacturing sector—a transition that the villagers were not equipped to cope with). About two-thirds of displaced household and unaffected household members had to change their occupation or industry compared to about 23% amongst the land acquired households. Spillover effects have also led to unaffected households losing some of their traditional livelihoods due to the setting up of FSEZ. However, a larger proportion of members from displaced households have found employment within the FSEZ, compared to members of land acquired households.

Female labour market participation was about six to seven times that of men. Unaffected household men and women reported higher labour market participation rates compared to both displaced and land acquired households. Tejani (2011) and Aggarwal (2007) report that industrialisation of rural villages results in increased feminisation of the workforce. Mansingh et al. (2012), similarly report large and increasing proportions of female employment across SEZs in India, with the FSEZ, Madras SEZ and Kandla SEZ reporting the highest proportions of female employment.

Empirical Results

Profiles of the Active Labour Market Participants

Affected household members were less likely to participate in the labour market. The probit outcomes, using a dummy dependent for labour force participation, in Table 7.1 show that working-age members from displaced households were about 4–5 percentage points (or 19–24%) less likely to

Table 7.1 Profiles of active labour market participants

Dependent variable = Labour force participation rate	*(1)*	*(2)*	*(3)*	*(4)*
Displaced	−0.126**	−0.100*	−0.106*	−0.054
	(−0.048)	(−0.035)	(−0.037)	(−0.018)
Land taken	−0.223***	−0.195**	−0.178**	−0.171**
	(−0.082)	(−0.073)	(−0.067)	(−0.064)
Female (Yes = 1)	−2.041***	−2.077***	−2.088***	−2.029***
	(−0.661)	(−0.673)	(−0.675)	(−0.661)
Log(PCHHE)			0.200***	0.202***
			(0.078)	(0.079)
Displaced × Female				−0.128
				(−0.048)
Individual controls	✓	✓	✓	✓
Household controls		✓	✓	✓
Constant	−1.146***	−0.873**	−2.730***	−2.767***
N	3292	3292	3292	3292

Note: ***$p < 0.01$, **$p < 0.05$, *$p < 0.1$. Marginal fixed effects are reported in parentheses. Robust standard errors clustered at the household. The sample comprises the working-age population, defined as those between the ages of 15 and 65 inclusive. Individual-level controls include the following covariates: natural logarithm of the age, dummies for level of education and dummies for marital status. Household-level controls include the following covariates: household size, dummy for split households, number of children, ethnicity/religion and the number of male and female working adult members. PCHHE—Per capita household expenditure

take part in the labour market compared to working-age members from unaffected households. Labour force participation rate, here, is estimated using a dummy dependent variable that takes a value of 1 if the individual is currently employed and 0 otherwise. Similarly, working-age members from land acquired households were about 7–8 percentage points (or 35–40%) less likely to participate in the labour market compared to working-age members of unaffected households. Therefore, there is some evidence to indicate that affected households were disadvantaged due to the setting up of FSEZ. However, introduction of additional controls at the household level weakens the power and size of the coefficient. Working-age members of households with higher per capita consumption were more likely to participate in the workforce. Female labour force participation, however, is lower across all model specifications. Inclusion of the per capita household expenditure (PCHHE) as a control in the last two specifications indicates a strong positive relationship between income (proxied by PCHHE) and labour force participation. The PCHHE and labour force participation are likely to be endogenous because of reverse causality and simultaneity; however, inclusion of PCHHE does not change the baseline results, and therefore the main findings are likely to hold. All models use standard individual and household-specific controls, discussed in the notes to Table 7.1. The inclusion of an interaction between household categories and gender shows that women from displaced households are less likely to participate in the labour market; however, the coefficient is not statistically significant.

The next section of the chapter explores if livelihood compensation in terms of promised jobs reached the affected household members.

Profiles of the FSEZ Employees

Displaced household members who were working were more likely to work within FSEZ. This is an important concern, given the job promises made to affected households at the time and provisions in current state legislation which explicitly identify livelihood-based compensation. The outcomes in the first four columns of Table 7.2 are based on the sample of working-age members and those in the last two columns are based only on active labour market participants. The marginal effects seem to indicate that members from displaced households were about 3–4 percentage points (or 11–15%) more likely to be employed within the FSEZ compared to those of unaffected households. Members from land acquired households, however, were about 23% less likely to be employed within the FSEZ. Since land acquired households used to be agricultural land owners who enjoyed higher income, it is possible that a higher reservation wage disfavours employment within the FSEZ (see the following sections for results on returns to education and experience at the FSEZ). Once household characteristics are controlled, these effects become statistically insignificant. Once household welfare indicators and interactions between genders and being displaced are introduced, the results indicate that women from displaced households were about 5 percentage points less likely to participate in the FSEZ. Inclusion of PCHHE as a control, results of which are reported in columns three and four, shows a strong positive relationship between income and employment at FSEZ. Again, despite the endogeneity in including PCHHE, given that the inclusion of PCHHE does not change the baseline results, the main findings are unlikely to change. However, when the sample is restricted to active labour market participants, displaced household

Table 7.2 Profiles of FSEZ workers among active labour market participants

Sample	*Working-age population*				*Active labour*	
Dependent variable = Work in FSEZ	*(1)*	*(2)*	*(3)*	*(4)*	*(5)*	*(6)*
Displaced	0.171**	0.109	0.092	0.224**	0.250**	0.314***
	(0.028)	(0.018)	(0.015)	(0.036)	(0.079)	(0.099)
Land taken	−0.412***	−0.230	−0.211	−0.200	−0.178	−0.167
	(−0.057)	(−0.033)	(−0.030)	(−0.029)	(−0.054)	(−0.051)
Female (Yes = 1)	−0.825***	−0.872***	−0.882***	−0.671***	0.591***	0.778***
	(−0.136)	(−0.136)	(−0.137)	(−0.103)	(0.211)	(0.282)
Log(PCHHE)			0.222***	0.222***		
			(0.034)	(0.034)		
Displaced × Female				−0.407***		−0.402*
				(−0.054)		(−0.110)
Individual controls	✓	✓	✓	✓	✓	✓
Household controls		✓	✓	✓	✓	✓
Constant	−0.847**	0.187	−1.850**	−1.916**	2.311***	2.312***
N	3292	3292	3292	1377	1377	1377

Note: ***$p < 0.01$, **$p < 0.05$, *$p < 0.1$. Marginal fixed effects are reported in parentheses. Robust standard errors clustered at the household. For columns (1) to (4), the sample comprises the working-age population, defined as those between the ages of 15 and 65 inclusive; for columns (5) to (6), the sample comprises those actively participating in the labour market and the corresponding dependent variable being the percentage of active labour market participants working within FSEZ. Individual-level controls include the following covariates: natural logarithm of the age, dummies for level of education and dummies for marital status. Household-level controls include the following covariates: household size, dummy for split households, number of children, ethnicity/religion and the number of male and female working adult members. PCHHE—Per capita household expenditure

members were about 17–24% more likely to be employed within the FSEZ. Women from displaced households, however, were statistically no different in employability within the FSEZ.

Determinants of Wages

This section explores the estimation of the Mincerian wage regression model (Mincer 1974) to identify returns to education, experience and other factors associated with earnings. Returns to education and experience have been widely studied by labour economists using the human capital theory (Mincer 1958; Becker 2008). In Mincer's proposed earnings function, the natural logarithm of earnings or wage is taken as a function of the key determinants of the accumulated human capital. The latter is identified in the function by a linear term for schooling and a linear and quadratic term for labour market experience. Therefore, the following function is estimated:

$$\ln(W_i) = \beta_0 + \beta_1(\text{Schooling})_i + \beta_3 \ln(\text{Experience})_i + \beta_4 \ln(\text{Experience})_i^2 + e_I \quad (7.1)$$

Ordinary Least Squares (OLS) estimations of Eq. (7.1) using the sample of active labour market participants are reported in Table 7.3. Log of annual wages is the dependent variable in all models. The results reported in Table 7.3 show pooled runs comprising three household categories. The first column shows outcomes with only individual-level controls, while the outcome in the second column also account for household-level controls. In the third and fourth columns, OLS outcomes controlling for occupation and industry fixed effects are shown, and the fifth column shows result after controlling for individual, household, occupation and industry-specific controls. Interactive terms with gender, education and being displaced are also used; the outcomes of these estimates are shown in the last two columns.

Two important findings stand out from the regression outcomes, as shown in Table 7.3. First, on average, land acquired households earned 12–20% higher wages than unaffected households. This outcome is consistent across all models and is statistically significant. The margin is higher when industry-specific effects are controlled. Second, average earnings of women were 11–19% lower than that of men. This effect is statistically significant across all specifications. The results also indicate that members of displaced households, on average, earn more than members of unaffected households; however, the coefficients are not statistically significant. Outcomes from models with interactive terms are particularly discouraging for women from displaced households. Returns to secondary and higher education are positive; the statistically significant coefficients indicate a range of 9–30% increase in wages with education levels. Returns to experience are positive, but the estimated coefficient is statistically insignificant. Surprisingly, the education premium for displaced household members yields a negative return, as shown in the last column of Table 7.3 (although this effect is not statistically significant). This is because members of displaced households earn less if they work at the FSEZ, while

Table 7.3 Determinants of wages for all employees

Dependent variable = log(wage)	*(1)*	*(2)*	*(3)*	*(4)*	*(5)*	*(6)*	*(7)*
Displaced	0.009	0.031	0.016	0.014	0.049	0.057*	0.014
Land taken	0.173***	0.121**	0.164***	0.198***	0.109*	0.128**	0.119*
Work in FSEZ (Yes = 1)	−0.016	−0.022	−0.005	0.013	0.002	−0.025	−0.020
Female (Yes = 1)	−0.170***	−0.193***	−0.175***	−0.174***	−0.189***	−0.112**	−0.194***
Log(Experience)	0.072	−0.025	−0.024	−0.032	−0.028	−0.026	−0.027
Log(Experience)^2	0.000	0.000	0.000	0.000	0.000	0.000	0.000
Primary (Yes = 1)	0.021	0.020	0.010	−0.000	−0.006	0.016	−0.019
Secondary (Yes = 1)	0.095*	0.094*	0.070	0.054	0.048	0.089*	0.091
Higher (Yes = 1)	0.225*	0.238*	0.109	0.078	0.045	0.230*	0.300*
Displaced × Female						−0.185**	
Displaced × Primary							0.081
Displaced × Secondary							0.003
Displaced × Higher							−0.173
Occupation category			✓		✓		
Industry category				✓	✓		
Household controls		✓			✓	✓	✓
Constant	7.363***	7.719***	8.243***	7.628***	8.575***	7.659***	7.668***
R^2	0.126	0.137	0.181	0.165	0.214	0.140	0.140
N	1376	1376	1376	1376	1376	1376	1376

Note: ***$p < 0.01$, **$p < 0.05$, *$p < 0.1$. Robust standard errors clustered at the household. The sample comprises those actively participating in the labour market. At the individual level, we also control for age and marital status. Household-level controls include the following covariates: household size, dummy for split households, number of children, ethnicity/religion and the number of male and female working adult members

members of other household groups earn more if they work at the FSEZ. These results provide mixed evidence on the effect of SEZs on improving livelihood incomes and reducing gender wage gaps.

Determinants of Wages, for FSEZ Employees

Among the FSEZ employees, members of displaced households and land acquired households are the lowest and the highest earners, respectively. The estimates shown in Table 7.4 use a restricted sample of only household members working within the FSEZ. The Mincerian wage regression models explain 16–23% variation in wages for the FSEZ employees. In most of the models, the outcome is statistically significant for FSEZ employees from displaced households. Female employees in FSEZ from displaced households earn particularly less. Returns to education and experience show mixed outcomes but are not statistically significant.

To summarise, the empirical results indicate strong positive returns to education among all three categories of households, with the exception for those working within the FSEZ. The findings depict three distinctive features of FSEZ employees in the context of wage determination. First, returns to education and experience for FSEZ employees show a lower, statistically insignificant and often negative coefficient compared to what could be found in general. Second, the magnitude of the gender wage gap against women is 5–10% higher for the FSEZ employees. Third, secondary and higher education premium for members of displaced households yield positive returns when working at the FSEZ, as opposed to showing a negative relationship in general. One possible explanation for insignificant returns to education and experience, based on FGDs, is that the nature of work within the FSEZ is unskilled, and did not necessarily favour those with experience. These results question the conventional wisdom that SEZs improve education and gender parity.

Education, Employment and the Gender Gap

Labour force participation (LFP) rates are lower for women and affected households in our sample. There is also evidence of heterogeneity across age, which considers two cohorts: 15–35 years as young group and 45–65 years as old group. For displaced households, the lower LFP rate of members in the old group (only 11% as compared to 20% for land acquired and 26% for unaffected household) might explain the overall lower LFP

Table 7.4 Determinants of wages for FSEZ employees

Dependent variable = log(wage)	*(1)*	*(2)*	*(3)*	*(4)*	*(5)*	*(6)*	*(7)*
Displaced	−0.118**	−0.116**	−0.116**	−0.121**	−0.114*	−0.051	−0.118
Land taken	0.102	0.084	0.079	0.104*	0.071	0.089	0.082
Female (Yes = 1)	−0.275***	−0.256***	−0.263***	−0.265***	−0.238***	−0.122**	−0.254***
Log(Age)	1.074**	1.026*	0.871	1.037*	0.711	0.863	1.011*
Log(Experience)	−0.035	−0.034	−0.027	−0.034	−0.021	−0.025	−0.033
Log(Experience)^2	0.000	0.000	0.000	0.000	0.000	0.000	0.000
Primary (Yes = 1)	−0.109	−0.107	−0.088	−0.115	−0.084	−0.106	−0.088
Secondary (Yes = 1)	−0.069	−0.070	−0.056	−0.074	−0.043	−0.077	−0.084
Higher (Yes = 1)	−0.145	−0.158	−0.139	−0.158	−0.118	−0.138	−0.167
Displaced × Female						−0.274*	
Displaced × Primary							−0.031
Displaced × Secondary							0.026
Displaced × Higher							0.024
Occupation category			✓		✓		
Industry category				✓	✓		
Household controls		✓			✓	✓	✓
Constant	7.830***	7.949***	8.721***	7.786***	9.014***	8.364***	7.985***
R^2	0.165	0.188	0.199	0.182	0.226	0.204	0.188
N	367	367	367	367	367	367	367

Note: ***$p < 0.01$, **$p < 0.05$, *$p < 0.1$. Robust standard errors are reported in parentheses. Robust standard errors clustered at the household. The sample comprises those who were currently employed within the FSEZ. At the individual level, we also control for age and marital status. Household-level controls include the following covariates: household size, dummy for split households, number of children, ethnicity/religion and the number of male and female working adult members rate. Similarly, for land acquired households, lower LFP rate among the young group (stands at 54%, as compared to 63% for displaced and 67% for unaffected households) could similarly explain the overall lower LFP rate, compared to unaffected households. Findings from FGDs suggest that one reason for lower LFP rates for land acquired households could be a large number of teenagers and young adults from these households pursuing higher studies.

FSEZ and the Informal Labour Market

Among displaced households, about 36% and 25% of active male employees from young and old cohorts, respectively, were attached to FSEZ-related employment; this is higher compared to members from other household groups. The female participation rate within FSEZ is similar between displaced and unaffected households. Land acquired households were less likely to work in the FSEZ. The nature of employment at the FSEZ remains predominantly informal. Such informal positions are 'contracted-out' through 'contractors', a term locally used to refer to middlemen who act as agents recruiting on behalf of firms operating with the FSEZ. Mansingh et al. (2012) and Suchitra (2007) identify similar informal recruitment and labour management arrangements being present in other SEZs around the country. Contractors have a regional presence; this defines their area of operation and expertise. The contractors only source for unskilled labour (casual labour) and are not awarded contracts for 'desk jobs' (a local term used to refer to semi-skilled and skilled jobs). FGDs indicate that contractors are primarily used by the companies to pass on the responsibility and take no legal binding on the employment of casual labour. The contracting of work to middlemen also allows firms' flexibility on the staffing, as one villager noted 'hiring and firing if and when they want'.

Wages and the Informal Labour Market

Figure 7.2 plots predicted log wages across years of schooling by household categories depicting some interesting trends. For all employees, returns to education show similar trends across years of education. For FSEZ employees, the returns are lower for displaced households; this outcome is driven by female employees. There is also a higher gender wage gap in returns to education for female members from displaced households

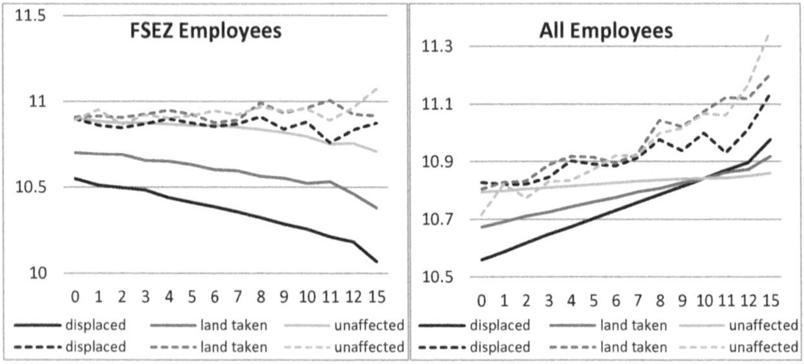

Fig. 7.2 Predicted log wages across years of schooling by household categories. Note: The x-axis represents the years of schooling, while the y-axis represents the natural logarithm of wages. The solid lines represent female employees and the dotted lines represent male employees. Source: Authors' own calculations

employed in the FSEZ. Overall, for all the FSEZ employees, the predicted log wages are constant across the years of schooling, whereas there is a positive trend for those employed elsewhere. Based on findings in Mansingh et al. (2012), this is partly driven by lower wages (often at the minimum wage rate) being paid to members employed within the zones. These findings indicate that SEZs-led structural transformation do not necessarily induce better returns to education.

The informal labour market discussed in the previous section may also partly explain the lower wages within the FSEZ. FGDs indicate that the number of such 'contractors' was proportionally lower in Highland (village with displaced households), creating an oligopolistic market for work contracts with unfavourable wage terms for women from displaced households. The general negative trend in wages across years of schooling may also be explained using the same 'contractor' system. FGDs reveal that the contractors favour younger, less educated women when recruiting on behalf of the FSEZ firms.

Gender Wage Disparity

Gender wage disparities are higher in the FSEZ than elsewhere. Estimates in Table 7.3 showed that women earn about 17% less compared to men, after controlling for education and experience; this is significantly lower than the national statistics identified in Mahajan and Ramaswami (2015). However, the magnitude of gender wage gap against women is 5–10% higher for FSEZ employees. Descriptive statistics from our data indicate that there is a significant gender gap in the wages, especially among older affected household members. Land acquired household members seemed to on average earn more than displaced and unaffected household members across both gender groups. Given that the FSEZ participation rate for land acquired households was lower, the statistics seem to confirm that FSEZ wage rates may be lower than wage rates elsewhere. There is also evidence of gender wage and education gaps narrowing, however, the gap still persists and is significantly large among FSEZ employees. The fall in gender gap for wages is large for land acquired households, but there is a larger drop in gender gap for schooling for displaced households and unaffected households. Therefore, there is some inconclusive evidence that gender wage gaps have reduced over time partly due to narrowing gender gaps in education.

Endogeneity and Pre-FSEZ Characteristics

One concern with the OLS and probit estimates, that form the main results, is that they may be driven by pre-FSEZ differences among household groups and not necessarily by the setting up of the FSEZ. If such differences are correlated with current labour market participation, employability and wage bargaining, the baseline results are likely to be biased. To address this concern, the educational attainment of *elders*—those who are at least 45 years of age (i.e. at least 15 years at the time of the setting up of FSEZ)—is used as a proxy for wealth and educational attainment of their children (the current generation), by extent livelihood choices and returns to education. The threshold of 15 years of age was used to evade any reverse causality in the educational attainment of these individuals by the setting up of the FSEZ. In India, education is compulsory until the age of 14; ILO and UNICEF define child labour as those between the ages of 5–14. In lieu of these age definitions, 15 years of age at the time of setting up of FSEZ is used to define an *elder*. Psacharopoulos (1994) and Card (2001) show that returns to education are significantly larger for low-income households. Shea (2000) and Grawe and Mulligan (2002) show that, while intergenerational wealth transfer may have negligible effect on educational attainment, in general, in lower income families, these may be significant and large. Using these two narratives, the education attainment of those aged 45 and above, across these three categories of households, is compared to identify pre-FSEZ differences among them. Descriptive statistics indicate that educational attainment across elders in the three categories of households is largely heterogeneous and is highly positively correlated with the average educational attainment of household members now.

All baseline estimations were therefore re-estimated, additionally controlling for the educational attainment of elders in the households. Results for labour force participation and work in the FSEZ are not significantly different due to the inclusion of pre-FSEZ educational attainment. However, in the Mincerian specifications, wage differences across household categories are no longer statistically significant. These results suggest that current differences in returns to education may be more attributable to intergenerational knowledge and income transfers rather than the FSEZ. The results also indicate that less educated members from displaced households draw better returns to education, and better-educated women from displaced households reap better returns to education than men. These findings indicate that controlling for pre-FSEZ educational levels, displaced households in fact benefit from the setting up of the FSEZ. These findings provide some credence to the benefits of SEZs-led structural transformation in rural areas.

Conclusion

This chapter addresses growing concerns over the process of rapid industrialisation and forcible land grabbing in India. At present, with over 400 new SEZs being approved in India, it is important that livelihood effects of SEZs are assessed in the long term to understand the cumulative effects of displacement, structural transformation and local labour market dynamics. It is imperative that industrialisation-based development policies result in a positive-sum game through achieving greater public good in setting up SEZs but also account for the private costs borne by individuals and households resulting from loss of land, livelihoods and accelerated structural transformation.

Results from this study indicate that the setting up of FSEZ, initially, had an adverse effect on the livelihoods of those who were displaced, but due to trickle down effects in the long term (Banerjee et al. 2012), the differences between affected and unaffected households were not statistically different. These results are evidence of a cumulative effect over a period of almost three decades. The results generally indicate that over long periods, adverse effects of displacement on the livelihoods of those affected may fade, and the setting up of SEZs may lead to an efficiency loss over equity gain. In the case of the FSEZ, this can be partly attributed to the existence of informal labour markets, higher female employment among affected households and the accompanying narrowing of gender wage and education gaps.

The main findings indicate the need for a sustainable policy framework to address SEZ-related displacement, livelihood and gender issues. Policy frameworks also require affirmative implementation spearheaded by local administrative units. For example, FGDs indicate that, while vocational training institutes were set up as part of the FSEZ to train and equip displaced household members with skills that benefit their employment within FSEZ, these institutes were abandoned within the first couple of years since FSEZ became operational. Moreover, the widespread informal recruitments within FSEZ create insecurity in wage and employment, and limit access to healthcare benefits, educational benefits for children and so on. Mansingh et al. (2012) and Suchitra (2007) provide evidence for the lack of collective bargaining by members working within the zones, coupled with a lack of adequate implementation of labour laws leading to labour exploitation. These warrant action from the state as part of a broader policy framework to ensure implementation of labour laws that protect affected workers within SEZs—particularly those who had been evicted for the setting up of the SEZ, and women.

References

Aggarwal, A. (2007). Impact of special economic zones on employment, poverty and human development. *Working Paper No. 194*. New Delhi: Indian Council for Research on International Economic Relations.

Banerjee, A., Duflo, E., & Qian, N. (2012). On the road: Access to transportation infrastructure and economic growth in China. *NBER Working Papers 17897*. Cambridge, MA: National Bureau of Economic Research.

Becker, G. S. (2008). *Human capital: A theoretical and empirical analysis, with special reference to education*. Chicago, IL: The University of Chicago Press.

Card, D. (2001). Estimating the return to schooling: Progress on some persistent econometric problems. *Econometrica, 69*(5), 1127–1160.

Grawe, N. D., & Mulligan, C. B. (2002). Economic interpretations of intergenerational correlations. *Journal of Economic Perspectives, 16*(3), 45–48.

Kundra, A. (2000). *The performance of India's export zones: A comparison with the Chinese approach*. New Delhi: Sage Publications.

Mahajan, K., & Ramaswami, B. (2015). Caste, female labour supply and the gender wage gap in India: Boserup revisited. *Discussion Paper 15-02*. New Delhi: Indian Statistical Institute.

Mansingh, P., Eluri, S., & Sreejesh, N. P. (2012). *Trade unions and special economic zones in India*. Geneva: International Labour Office.

Mincer, J. (1958). Investment in human capital and personal income distribution. *Journal of Political Economy, 66*(4), 281–302.

Mincer, J. (1974). *Schooling, experience, and earnings*. New York: National Bureau of Economic Research.

Mukherjee, A., Pal, P., Deb, S., Ray, S., & Goyal, T. M. (2016). *Special economic zones in India: Status, issues and potential*. New Delhi: Springer.

Nielsen, K. (2010). Contesting India's development? Industrialisation, land acquisition and protest in West Bengal. *Forum for Development Studies, 37*(2), 145–170.

Psacharopoulos, G. (1994). Returns to investment in education: A global update. *World Development, 22*(9), 1325–1343.

Roy, D. (Ed.). (2008). *Special economic zones: How special and how economic?* New Delhi: Hazards Centre.

Shalti Research Group. (2008). *SEZ in West Bengal: Growth trajectory impact*. Kolkata: Shalti Samiti.

Shea, J. (2000). Does parents' money matter? *Journal of Public Economics, 77*(2), 155–184.

Suchitra, S. (2007). SEZs: Economic or exploitation zones? *Infochange News and Features*. Retrieved December 17, 2014, from http://infochangeindia.org/agenda/cost-of-liberalisation/sezs-economic-or-exploitation-zones.html

Tejani, S. (2011). The gender dimension of special economic zones. In T. Farole & G. Akinci (Eds.), *Special economic zones: Progress, emerging challenges, and future directions* (pp. 247–282). Washington, DC: World Bank.

Wooldridge, J. M. (2009). *Introductory econometrics: A modern approach* (4th ed., p. 456). Boston, MA: Cengage Learning.

CHAPTER 8

Attitudes Towards Forced Displacement in West Bengal, India: To Move (Again) or Not?

Saumik Paul and Vengadeshvaran Sarma

Introduction

Development-induced displacement is widely abundant in the developing world (Gasper 2015; Bennett and McDowell 2012). Annually 10–15 million people are forced to relocate to carry on development projects (Gasper 2015). As Bennett and McDowell (2012) argue, in the developing world, most of those chosen for 'displacement' for development purposes come from those in the fringes of society—rather than those from the mainstream. Given that they are mostly poor, yield almost no political clout, and often at times do not have appropriate legal documents to their land, they have increasingly become vulnerable and susceptible to such large-scale 'land grabs' by the state. While Cernea (1988) and Gasper (2015) argue the importance of consultation and inclusive resettlement

S. Paul (✉)
Asian Development Bank Institute (ADBI), Tokyo, Japan
e-mail: spaul@adbi.org

V. Sarma
University of Nottingham, Semenyih, Malaysia

N. Yoshino, S. Paul (eds.), *Land Acquisition in Asia*,
https://doi.org/10.1007/978-981-13-6455-6_8

programmes, Garikapati (2005), Ghatak and Mookherjee (2011) and Ghatak and Mookherjee (2014) show that such consultations are likely to be hampered by two things: lack of collective consensus and higher reservation prices for land.

India in particular has witnessed several protests and hurdles in recent times in enforcing displacement for development agendas (see Sarkar 2007; Shah 2013). Since the late 1990s, post-liberalization India has witnessed an upsurge in demand for agricultural land for industrialization purposes (Searle 2010). This ostensibly brings farmers at the crossroads where they make a choice between relocation and stay put. Two factors obscure the choice. The majority of land in India is owned by small peasantry, often with unclear land titles (World Bank 2007). The state acquisition of land, based on the Land Acquisition Act of 1894, allows the use of *eminent domain* for public purposes. While the new Rehabilitation and Resettlement Act of 2013 replaced this by now requiring at least 80% consent from all affected parties, it may exert peer pressure on individual farmers facing the collective consensus of a community. Together, these factors obviate land owners from collective bargaining and they frequently part away their land despite refusal.

Why do individual land owners refuse to sell their land for industrialization purpose? Over the past decade, this question has received repeated attention from policymakers and scholars alike, as the conflict between farmers and the industrial sector over land revolved around special economic zones (SEZs). State-sponsored 'land grabs' have led to widespread social and political tensions in many parts of India (Sarkar 2007; Shah 2013). This has resulted in casualties and severe injuries to many protesters, eventually leading to the cancellation of proposed SEZs, as witnessed in Nandigram and Singur in West Bengal (Levien 2011).

As scholarly works indicate, myriad facets of life impinge on such relocation decisions for farmers. While under-compensation of arable land, dependence on agricultural land for livelihood, and non-inherited ownership contribute to this refusal (Ghatak et al. 2013), some owners also worry about potential damage to the local environment from SEZs and are sceptical of intergenerational benefits (Bardhan 2011). These studies provide useful evidence on owners' attitudes but they are mostly conducted in disputed SEZ sites (such as Singur), where farmers had already chosen not to relocate prior to surveys. Little is known on the farmer-owner attitude towards the prospects of a future move subject on an adequate compensation package. It is also pertinent to learn about how farmers, who had similar experiences in the past, perceive relocation. This is important for two reasons: (1) the expansion of existing SEZs and (2) reflection of forced displacement experience(s) and its effect in shaping future *voluntary* displacements. First, most relocated people in India, due to the setting up of SEZs, stay in the vicinity of those SEZs (Shalti Research Group 2008); as such, any expansion to the SEZs is likely to (re)affect such displaced and resettled populous. Second, having gone through forced displacement, the expectations and gap in asymmetry of information (based on experiential learning) are likely to be different for those with prior displacement experiences and those without. There has been a dearth of studies that study these aspects of displacement and we intend to fill this knowledge gap based on a self-administered household survey.

We use a novel dataset based on a household survey near the Falta Special Economic Zone (FSEZ) in West Bengal, India, conducted during May and June 2013. The FSEZ, set up in 1984, has been a relatively successful SEZ in West Bengal (Shalti Research Group 2008). Our data allow for comparison between those who were previously displaced (in 1984) and those who were not. We conjecture that previously affected households base their decisions of (re)displacement and compensation on previous experience(s)—including the quality and proportion of compensation received. Other households are likely to base their decisions on asymmetric information. This classification of the sample allows us to identify heterogeneity in attitudes towards displacement. In addition, we also study whether other household and demographic characteristics matter in the relocation decision.

Our data comprise of 1017 households: 462 who lost land and were displaced, 168 who lost land but were not displaced and 387 who were not directly affected. Households were asked to *Imagine that there will be a possible extension of the FSEZ in the coming 12 months and that your household will lie within the expansion area. Are you willing to move out with an adequate compensation package?* A number of questions on optimal compensation, legal notice period and reasons for not moving were also asked. Almost 16% of our sample showed preferences for land transfer; this includes 18% of those from unaffected households, followed by 16% and 8% from displaced and land-acquired households, respectively.

This chapter is structured as follows. In the section 'Setup of FSEZ, Displacement and Compensation Setup of FSEZ' we briefly recap the FSEZ background and outline the state compensation policy at the time to provide some contextual background. The section 'State-Approved

Compensation Policy' describes the survey design and provides some descriptive evidence. We present the empirical findings in the section 'Data' and present a discussion of these in the section 'Results'. The section 'Conclusion' provides our concluding remarks.

Setup of FSEZ, Displacement and Compensation

Setup of FSEZ

The Falta Special Economic Zone (FSEZ) was set up in 1984 by the Government of India with assistance from the state government of West Bengal. Evidence suggests that these villagers were forcibly evicted. Occupants of these villages were never consulted in prior regarding their preferences for resettlement or were not part of any consultation groups that devised compensation formulae (Shalti Research Group 2008; Aggarwal 2012; and our own focus group discussions [FGDs]). According to the Shalti foundation and our own focus group discussions, we learn that only a 'notice' was issued to the villagers informing them of the acquisition and a promise for adequate compensation. The notice supposedly read that the villagers should evacuate 'within 30 days' and failing such, 'forcible acquisition' would be carried out. The studies and our FGDs also suggest that the state actors responsible for the setting up of SEZs promised that 'one person from every household would be offered a job in the Free Trade Zone (FTZ)'. An additional 80 acres of land was also acquired from a third village—'Gopalpur'—to resettle affected villagers from the aforementioned two villages of Akalmegh and Uttar Simulberia. While most villagers from Akalmegh and Uttar Simulberia had to give up their residence and residential land, most villagers from Gopalpur had to give up their agricultural land for the settling of those affected villagers from Akalmegh and Uttar Simulberia. Similar to the experience of villagers from Akalmegh and Uttar Simulberia, villagers from Gopalpur also received a similar 'notice' and were never part of consultations prior to the acquisition of their land. Prior to the setup of FSEZ, most of these villagers (more than two-thirds) relied on agriculture, fishing and ferrying goods on boats (locally referred to as boating) for their livelihood. Since the acquisition of land for FSEZ and the setup of FSEZ, about 60% of the villagers have changed their occupation and/or industry—mostly from the primary industry to low-skilled jobs in manufacturing (Paul and Sarma 2013; Shalti Research Group 2008).

State-Approved Compensation Policy

The land for the Falta Special Economic Zone (FSEZ) was acquired under the Land Acquisition Act, 1894. The colonial-era act gave absolute power to the state for forcible acquisition of land for 'development' purposes and to promote 'national interests'.[1] The government's notification No. S.O. 782 (E) dated 25 October 1984 declared that land was to be acquired for the FSEZ. The land acquired was rich in alluvial soil and was very fertile. A huge variety of crops, vegetables and fruits were grown in the area. Despite the attraction of regular jobs and better standards of living, most affected villagers were unwilling to part with the lustier and green pastures that were home to them.

A committee was thus set up to look after the compensation and rehabilitation packages. The committee consisted of a number of people of from the locality and also the then district magistrate and a member of the Akalmegh Gram Panchayat. The committee promised five things: (1) a job per family; (2) housing with basic amenities; (3) free electricity to all households; (4) better standards of living for the evicted party; and (5) resettlement of evicted households in a model village.

The committee had also decided that the evicted parties would be resettled in a model village (presently known as Highland). In addition, the committee drafted a tabulation of the victimization categories, definition and conditions that qualify victims for each category. The categories were as follows: (1) those who lost residential land and residence and (2) those who lost agricultural land but not their residential land or residence. Those who lost their livelihoods as a result of the acquisition were not covered in any of these categories and as a result their loss of livelihood was never compensated for.

Data

Sampling

Our data and sampling follow from those discussed in Chap. 7. Our data therefore comprise of 1017 households from these three villages. 630 of these households were affected by the setup of FSEZ (462 households lost their residential land, residences and were forced to resettle elsewhere and 168 of these households lost agricultural land but did not have to relocate). An additional 387 unaffected households from the village of Nainan form our control group (refer to Table 8.1).

Table 8.1 Households categorized by village and category

Village	*Affected and displaced*	*Affected but not displaced*	*Unaffected*	*Total*
Highland	460	1	5	466
Gopalpur	0	107	52	159
Nainan	2	60	330	392
Total	462	168	387	1017

Source: Authors' own calculations

Compensation in Reality

Evidence suggests that compensation was hard to come by. Findings from a study by the Shalti Research Group (2008) indicate that the first batch of 10–12 relocated households was given Rs. 1200 (about USD 65 based on historical exchange rates) per household. The government and local Panchayat body decided to provide the evicted families land in the model village amounting to 4/6/8/10 decimal per household based on the previous holdings of land of the respective household before eviction. The Shalti foundation's study and our own focus group discussions indicate that there was no appropriate measurement of formulae for the compensation of the dwellings; the compensation was mostly given on the presumptions made by the officials.

Further anecdotes in the Shalti foundation's study indicate that the compensation for evicted villagers from Akalmegh and Simulberia was fixed at Rs. 5700/- per bigha of land. In addition, the people of Gopalpur, who lost their agricultural land for the purpose of resettlement, received very low amounts of compensation. Compensation paid to them in some instances was as low as Rs. 5/ (about USD 0.27 based on historical exchange rates) per bigha (Shalti Research Group 2008). Until recently, as late as 2007, most villagers had not received their compensation in full. Only in the latter parts of the decade (2008–2010) did they receive a backlogged compensation that was to be offset against compensation that they were promised (but the hand-out compensations were unadjusted for inflation/time value of money).[2] To further add to the quandary, the evicted villagers do not have deeds to their present residential plots to date.

As part of the acquisition and resettlement packages, those affected by the setup of FSEZ were promised both cash compensation and residences (where the residence was lost). The compensation package promised to the villagers was a function of their location of residence (i.e. Akalmegh/Simulberia v Gopalpur) and of their land and residence holding. Those who were displaced (from Akalmegh and Simulberia) irrespective of land and residence holdings were promised a new residence and residential plot in 'Highland'. In addition, they were also to be compensated in cash for any land or residence that they had to forego due to eviction. On the other hand villagers from Gopalpur were only compensated in cash for the loss of agricultural land—held in 'Highland'. Table 8.2 shows the rate of compensation received by the affected households for residential and agricultural land.[3]

A caveat with the compensation received by these villagers was that, despite the rates exhibited in the table (first row) being those agreed by the committee that was enacted to facilitate the setup of FSEZ—they were agreed in 1984, but paid only in 2008 (thus, not accounting for the time value of money). These politically motivated payouts, which coincided with parliamentary elections and in the immediate aftermath of the Nandigram troubles (see Sarkar 2007), failed to adjust compensation payments for inflation. Even accounting only for inflation (and ignoring appreciation of land value, which surpassed inflation by almost a factor of two), we still find that villagers were under-compensated by about five times their due compensation.

Displaced households, who lost residences, were severely under-compensated; more than two-thirds of the villagers received compensations of less than Rs. 5000 (approximately USD 100, based on 2008 exchange rates). Given that in West Bengal the minimum wage laws at 2013 stipulated the minimum wage to be more than Rs. 5000/- per month (while this is the monthly wage floor, based on occupational category, this is likely to be much higher), villagers pointed out that the received compensation was not significantly different from their regular

Table 8.2 Loss of land/residence by category of household

Category	*Cultivable land (in* katta*)*	*Residential land (in* katta*)*	*Residence (in Rs.)*
Affected and displaced	54.57 (136)	10.41 (444)	43,568.29 (455)
Affected but not displaced	34.69 (166)	4.86 (21)	32,133.33 (55)
Total	43.65 (302)	10.16 (465)	43,203.35 (470)

Source: Authors' own calculations

Note: The numbers indicate means; the number of observations are indicated in parentheses

Table 8.3 Compensation of land by category of household

Proportion of compensation received	*Affected and displaced*	*Affected but not displaced*	*Total*
0	0.00	1.24	0.70
1–49%	4.84	3.72	4.2
50–79%	4.03	4.35	4.21
80%	90.32	77.64	83.16
100%	0.81	13.04	7.72

Source: Authors' own calculations

income and that this was fungible and did not represent an exogenous addition to their wealth.

The descriptive statistics from Table 8.3 indicates that, post-2008, a lion's share of households received compensation for 80% of their land, both residential and agricultural. Locally weighted polynomial regressions shed some light on the distribution of this compensation. The analysis shows that land compensation for displaced households depicted a bimodal distribution. In contrast to households who lost land for relocation, median households on the scale of the amount of land lost received the lowest compensation rate. This illustrates some heterogeneity in the receipt of compensation.

Similarly, even in the case of compensation for lost residences, less than 10% of the households received full compensation and locally weighted polynomial regressions. Table 8.3 again indicates greater loss for median households on the scale of the amount of land lost. Using this descriptive evidence and existing literature as guidelines, in the next section we produce some empirical results that look at the effect of these displacement and compensation experiences (or the lack of it—in the case of our control group) which affects the willingness to relocate in the future.

Results

Does Prior Displacement Matter?

We employ multivariate regression techniques to decipher the factors associated with the prospective relocation decision. Probit estimations of the willingness to move (a dummy variable, which takes the value of 'one' if willing to move, 'zero' otherwise) by the category of households (reported in Table 8.4) show that displaced households were 3.5% less likely to move compared to unaffected households. However, once we control for per capita household expenditure (PCHHE), the effect is insignificant statistically. There is also evidence to suggest that richer households are less likely to transfer land, while those who received larger compensation for their land previously are more willing to move now.

We also asked the following questions to those who were *willing to move*:

(a) *What is the optimal compensation package for land/plot acceptable to you? (in Rs./katta)*
(b) *What is the optimal compensation package for your residence acceptable to you? (in Rs.)*
(c) *What is the legal notice timeframe preferred (in months)?*

Using responses to these questions, ordinary least squares estimates predict that households with prior experience of displacement show lower expectations on the compensation rate compared to other households

Table 8.4 Willingness to transfer land and prior displacement

Dependent variable: Willingness to move			
	(1)	*(2)*	*(3)*
People who lost land and were displaced	−0.186* (−0.035)	−0.148 (−0.026)	
People who lost land but were not displaced	−0.070 (0.014)	0.020 (0.004)	
Log(PCHHE)		−0.614*** (−0.110)	
Log of compensation received for residence			0.017 (0.004)
Log of compensation received for agricultural plot			0.082** (0.018)
Controls	✓	✓	✓
Constant	−5.510***	0.316	−2.730***
Pseudo-R^2	0.093	0.121	0.115
N	1017	1017	630

Note: ***$p < 0.001$, **$p < 0.01$, and *$p < 0.05$. Marginal fixed effects are reported in parentheses. Controls include household size, the number of children, ethnicity, the number of male and female working adult members, and a dummy for split households and age, education, and marital status of the head of the household. The sample in column (3) comprises only displaced households. PCHHE: per capita household expenditure

Table 8.5 Requirements for voluntary movement

Dependent variables	*Log compensation rate for land*	*Log compensation rate for residence*	*Timeframe of notice*
	(1)	*(2)*	*(3)*
People who lost land and were displaced	−0.474***	−0.533***	−0.235
	(0.149)	(0.172)	(0.653)
People who lost land but were not displaced	−0.455***	0.239	1.333
	(0.188)	(0.173)	(0.932)
Log(PCHHE)	0.915***	0.402*	−0.854
	(0.151)	(0.221)	(0.790)
Controls	✓	✓	✓
Constant	1.671	5.934***	19.026**
R^2	0.358	0.443	0.105
N	154	155	159

Note: ***$p < 0.001$, **$p < 0.01$, and *$p < 0.05$. Robust standard errors are reported in parentheses. Column (3) is measured in months. Controls are identical to those used in Table 8.1. The sample sizes vary because the number of responses received are varied

(reported in Table 8.5). Richer households, measured by PCHHE, are more likely to ask for a higher compensation both for land and for residences; this could be due to a higher reservation income that they enjoy. On average, households with prior displacement experiences also exhibit a lower timeframe of notice requirements; however, the outcome is not statistically significant.

Do Property Rights and Social Values Matter?

We find that households who possess titles for agricultural land and residential land are more willing to move, by about 19% and 8%, respectively. This indicates the risk-averse nature of land owners, who fear inadequate compensation in the absence of land deeds.[4] We did not find evidence to suggest that the size of the land matters in the decision to move (for brevity, we do not report this result here, but it is available upon request). Respondents were asked for their opinion on whether people can be trusted in their community and whether political connections were the most important factor to succeed in life in their locality. The last two columns in Table 8.6 reveal that those who believe that people can be trusted are 12% less likely to move. On the contrary, those who believe that political connections are the key to

Table 8.6 Willingness to move, property rights and social values

Dependent variable: Willingness to move				
	(1)	*(2)*	*(3)*	*(4)*
Deed for agricultural land (Yes = 1)	0.783*** (0.189)			
Log(PCHHE)	−0.599*** (−0.103)	−0.550*** (−0.097)	−0.725*** (−0.162)	−0.721*** (−0.123)
Deed for residential plot (Yes = 1)		0.429*** (0.081)		
Trust			−0.796*** (−0.119)	
Political connection				0.915*** (0.233)
Controls	✓	✓	✓	
Constant	0.395	−0.546	1.552	0.876
Pseudo-R^2	0.133	0.134	0.180	0.153
N	1017	1017	1017	1017

Note: $^{***}p < 0.001$, $^{**}p < 0.01$, and $^{*}p < 0.05$. Marginal fixed effects are reported in parentheses. Controls are identical to those used in Table 8.5

success are 23% more likely to transfer land and move out. These indicate that societal capital and positive attitudes towards success make one less likely to move out. Household income, occupation, industry and wages of household heads do not affect the land transfer decision; we do not show these outcomes for brevity and they are available upon request.

How Important Are Livelihood Choices?

We find that households who had at least one household member working in FSEZ were about 5% more likely, on average, to be willing to relocate. This indicates a general frustration with the livelihood options and prevailing working conditions within the FSEZ, we elaborate on this further in the following section. The result is statistically significant regardless of whether and what covariates we control for. The results also suggest that unskilled workers were about 5% more willing to relocate, on average, than semi-skilled, skilled or professional workers. Again this effect is statistically significant; however, the marginal fixed effects do not present a relatively large coefficient.

Our results point to some interesting nuances in the compensation and relocation debate. We explain these through three factors.

(a) *The ideal resettlement package*: Our evidence shows that prior displacement experience, whether directly or indirectly, reduces the reservation price of land. Contrary to Garikapati's (2005) and Ghatak and Mookherjee's (2011) assertions that when consulted on relocation farmers are likely to exaggerate the price of land, we find that their optimal compensation package was on par with market rates (the mean of Rs. 100,452 per *katta* from our survey was roughly on par or just below the market average of Rs. 110,000). These indicate that prior displacement experiences—especially adverse ones—lower expectations on compensation from future displacement exercises.
(b) *A sense of belonging to the locality*: Similar to findings of Lam et al. (2016), we find that forced relocations deteriorate social cohesion and a sense of belonging among communities. About one-third of the respondents preferred to leave their current locality. Of the two-thirds who preferred to remain in the locality, 77% of them wanted to remain close to the FSEZ due to familiarity with livelihood options, only 14% wanted to stay back due to communal ties or bonding with the environment.
(c) *Employment at FSEZ*: The nature of employment at FSEZ remains predominantly informal. Such informal positions are 'contracted out' through 'contractors'—a term locally used to refer to middlemen who act as agents recruiting on behalf of FSEZ companies. ILO (2012) and Suchitra (2007) identify similar informal recruitment and labour management arrangements present in other SEZs around the country. Contractors have a regional/locale presence and bias, it defines their area of operation and expertise. The contractors only source for unskilled labour (casual labour) and are not awarded contracts for 'desk jobs' (a local term used to refer to semi-skilled and skilled jobs). Based on the accounts of locals, contractors are primarily used by companies to pass on responsibility and take no legal binding on the employment of casual labour. The contracting of work to middlemen also allows companies flexibility on the staffing, as one villager noted 'hiring and firing if and when they want'. This, coupled with infrequent job availabilities at FSEZ (predominantly for low-skilled workers), is likely to drive

the positive coefficient for willingness to move in Table 8.6. Such disgruntled workers are more likely to prefer to a newer location *conditional on a better package of employment* as evident in their responses to the survey.

Conclusion

We find that prior displacement experience, wealth, property rights and attitudes matter. While possession of property rights makes one more willing to transfer land, wealth and prior experience of land eviction dampen it. A positive attitude towards life and the presence of strong social capital also make one more likely to refuse such an offer. However, a positive association with compensation receipts in the past favours another move out, given an optimum compensation package. Expectations, in terms of compensation rates, are higher for those with no prior experience of receiving compensation—perhaps indicating an upward bias due to reliance on asymmetrical information. Overall our results indicate that consultation with villagers prior to a displacement may be beneficial for both those to be relocated and the state. Our results also indicate that contrary to findings in previous literature, villagers may in fact have realistic expectations on land compensation and resettlement packages. While there are heterogeneities in the expectations, eliciting preference for resettlement prior to such an experience may not be too costly or lead to high reservation prices among villagers.

Notes

1. Due to the populist revolt, as discussed in the introduction, the act has now been replaced with a more populist act (the differences between the colonial era-act and its 2013 successor are compared in the sub-section 'Compensation in Reality'.
2. These were indicated to us by participants of focus group discussions and local Panchayat officials.
3. Given that the Government of India enacted a new act that governs land acquisition (and also addresses resettlement), our findings are more reflective and should be contextualized within the framework of the Land Acquisition Act 1894. A comparison of the new land bill and the Land Acquisition Act 1894 demonstrates that while those affected by the setup of FSEZs received provisions not availed in the Land Acquisition Act of 1894,

they were however disadvantaged in comparison to provisions made in the recent 2013 act. Despite criticisms of increasing bureaucratic red tape in the acquisition of land by the private sector, the new act extends provisions that safeguard the interests of land owners and those dependent on that land, especially via requiring overwhelming consent among those potentially affected.

4. In most instances of state-led evictions, property titles are a pre-requisite for compensation (see Shalti Research Group 2008).

References

Aggarwal, A. (2012). *Social and economic impact of SEZs in India*. New Delhi: OUP Catalogue, Oxford University Press.

Bardhan, P. (2011, April). Land acquisition: Currently a major stumbling block for development policy. *Development Outreach*. World Bank.

Bennett, O., & McDowell, C. (2012). *Displaced: The human cost of development and resettlement*. New York: Palgrave Macmillan.

Cernea, M. M. (1988). Involuntary resettlement in development projects: Policy guidelines in World Bank-financed projects. *World Bank Technical Paper No. 80*. Washington, DC: World Bank.

Garikapati, S. (2005). Consulting the development-displaced regarding their resettlement: Is there a way? *Journal of Refugee Studies, 18*(3), 340–361.

Gasper, D. (2015). The ethics of economic development and human displacement. In G. F. DeMartino & D. N. McCloskey (Eds.), *The Oxford handbook of professional economic ethics*. Oxford: Oxford University Press.

Ghatak, M., Mitra, S., Mookherjee, D., & Nath, A. (2013). Land acquisition and compensation in Singur: What really happened? *Economic and Political Weekly, 48*(21), 32–44.

Ghatak, M., & Mookherjee, D. (2011). Land acquisition for industrialization and compensation of displaced farmers. *Working Paper No. 2011, 35*. World Institute for Development Economics Research.

Ghatak, M., & Mookherjee, D. (2014). Land acquisition for industrialization and compensation of displaced farmers. *Journal of Development Economics, 110*, 303–312.

International Labour Organization (ILO). (2012). *Trade unions and special economic zones in India*. Geneva: International Labour Office.

Lam, C., Paul, S., & Sarma, V. (2016). Reversal of fortune? The long-term effect of conservation-led displacement in Nepal. *Oxford Development Studies, 44*(4), 401–419.

Levien, M. (2011). Special economic zones and accumulation dispossession in India. *Journal of Agrarian Change, 11*(4), 454–483.

Paul, S., & Sarma V. (2013). The livelihood effects of industrialization on displaced households: Evidence from Falta Special Economic Zone, West Bengal. *ZEW Discussion Paper No. 13-083*, Centre for European Economic Research.

Sarkar, A. (2007). Development and displacement: Land acquisition in West Bengal. *Economic and Political Weekly, 42*(16), 1435–11442.

Searle, R. (2010). *Making space for capital: The production of global landscapes in contemporary India*. Unpublished dissertation, University of Pennsylvania.

Shah, A. (2013). Mainstreaming or marginalisation?: Evidence from special economic zones in Gujarat. *Economic and Political Weekly, 48*(41), 55–61.

Shalti Research Group. (2008). *SEZ in West Bengal: Growth trajectory impact.* Kolkata, India: Shalti Samiti.

Suchitra, S. (2007). SEZs: Economic or exploitation zones?. *InfoChange News & Features.* Retrieved October 17, 2015, from http://infochangeindia.org/agenda/cost-of-liberalisation/sezs-economic-or-exploitation-zones.html

World Bank. (2007). *India: Land policies for growth and poverty reduction.* New Delhi: Oxford University Press.

PART III

Way Forward: Policy Measures

CHAPTER 9

Land Trust to Facilitate Development Through Land Transfer

Naoyuki Yoshino, Saumik Paul, Vengadeshvaran Sarma, and Saloni Lakhia

Introduction

Development has almost always led to relocation, creating an entirely new structure of possession and proprietary rights. It is indeed true that whenever this happens, the locals who are systematically dispossessed are in a weak condition to resist it. Various instrumentalities ranging from sovereign rights on land to a promise of fair compensation have been applied to ensure a smooth transition of rights and ownership.

For the state too, it is an inevitability incongruity and a policy praxis marred with conflicting priorities and uncertain outcomes.

Since the late 1980s, human displacement as a conservation strategy has been commonly practised, particularly in developing countries despite the increasing criticism by social scientists that this represents a severe human rights violation. They argue that protected areas have most often

N. Yoshino • S. Paul (✉) • S. Lakhia
Asian Development Bank Institute (ADBI), Tokyo, Japan
e-mail: nyoshino@adbi.org; spaul@adbi.org; slakhia@adbi.org

V. Sarma
University of Nottingham, Semenyih, Malaysia

N. Yoshino, S. Paul (eds.), *Land Acquisition in Asia*,
https://doi.org/10.1007/978-981-13-6455-6_9

been created in regions where the most socially, economically and politically vulnerable groups reside (Agrawal and Redford 2009). In other words, conservation-induced displacement may seriously damage the traditional yet powerless social groups. Scholars point out that fundamental features of displacement involve not just loss of place and material assets but also loss of social ties, identity and meaning.

Below mentioned are certain examples within Asia that further study the development-displacement juxtaposition.

Nepal

Many studies show that since the 1950s, Nepal has experienced rapid transformation in land ownership from the indigenous economically marginalized groups who had the weakest political standing to the more powerful immigrant groups. It also closely corresponds to the local socio-economic context, particularly after the large number of immigrants led to disruptive social conflicts between indigenous and migrant groups. The backdrop of this socio-economic upheaval involved a displaced indigenous group, the *Rana Tharus* (hereafter referred to as *Ranas*) in the western-most district of Kanchanpur. They experienced a large-scale displacement due to the expansion of the Shuklaphanta Wildlife Reserve (hereafter referred to as the Park) in 2001. The Nepalese government carried out a land-based resettlement scheme. It was designed on the principle that all displaced families should be given cultivable land, which they lost previously due to the extension of the wildlife reserve (Bhattarai 2001, p. 270).

While quantitative evidence suggests that by and large the land-based compensation policy failed to prevent impoverishment in the *Rana* society, it does not provide much evidence on the social injustice that *Ranas* have been facing over a long period of time. The *Ranas*, like many traditional societies, failed to perceive the modern concept of land ownership as an exercise in land registration documents. For them, the concept of land ownership was more about actual land use practices. Guneratne (2002) explains that the concept of obtaining legal land documents to secure ownership does not exist among many tribal or ethnic communities, particularly those from the lowland Tarai region of Kanchanpur.

Malaysia

In recent years, concerns over property rights, displacement and the welfare of indigenous communities have centred in political and media landscapes—including the *Orang Asli* of Malaysia.

For instance, the construction of industrial parks, dams and even highways has displaced many indigenous *Orang Asli* (indigenous) communities across Malaysia. An estimated 76 *Temuan* families (one of the largest indigenous tribes) were displaced from their homes and ancestral lands around the areas of Sungai Selangor and Sungai Gerachi for the construction of the largest dam in Malaysia—the Sungai Selangor Dam. Of these, 37 families (160 persons) were located in Kg. Gerachi near Sungai Gerachi and 39 families (179 persons) in Kg. Pertak, 6 km upstream on Sungai Selangor.

Given the distance from the location of displacement and the difficult terrain that has to be endured, most socio-economic indicators point to a relatively adverse scenario for families that were moved to Kg. Gerachi compared to Kg. Pertak. Land ownership was almost consistent across both village communities but those in Kg. Gerachi did not possess land titles for their property compared to about 72% of those in Kg. Pertak who reported having deeds for their land. Financial literacy was higher among residents in Kg. Pertak compared to those in Kg. Gerachi. The government compensation policies in terms of providing agricultural land and trees to those in Kg. Gerachi and providing a fixed income for those in Kg. Pertak have been successful in averting deprivation among communities in these resettled villages. However, the afflictive location of Kg. Gerachi relative to Kg. Pertak had made residents of Kg. Gerachi perform adversely across multiple socio-economic indicators—including education, labour market participation and per capita household expenditure (see Chap. 5 for more details). These findings indicate the need for further attention in terms of policy and research, especially comparisons of other *Temuan* and *non-Temuan Orang Asli* groups who have been resettled.

India

Special economic zones (SEZs) have, over the past decade, become the epicentres of conflict between farmers and the industrial sector over land, in India. The majority of land is owned by small peasantry, often with unclear land titles. State-sponsored 'land grabs' have led to widespread social and political tensions in many parts of India. This has resulted in casualties and severe injuries to many protesters, eventually leading to the cancellation of the proposed special economic zones, as witnessed in Nandigram and Singur in the state of West Bengal. The Falta Special Economic Zone, set up in 1984, has been a relatively successful SEZ in West Bengal. With regard to attitude towards relocation, it was found that while possession of property rights makes one more willing to transfer land, wealth and prior experience of land eviction dampen it. A positive attitude towards life and the presence of strong social capital also make one refuse such an offer. However, a positive association with compensation receipts in the past favours another move out, with an optimum compensation package.

Land rights and sovereignty interact in a variety of manners. History of the land, its legal system, ownership patterns and even local culture determine the dynamics of transition. The transfer of ownership, especially where the driving forces are not intrinsic, is bound to lead to strife; the very purpose of law is to ensure a smooth transition while safeguarding the rights of those affected. Below mentioned are two different approaches of land development. Japan has clear legal provisions, with a new innovative process of a land trust structure that is proposed in the section 'Land Trust Structure'.

India, however, is grappling with multiple land disputes and is also trapped in legal uncertainty with regard to its land acquisition laws. In the absence of a clear central law, the states have been given the power to create their own land laws in the interim period. While the section 'Land Trust Law, Japan' focuses on land laws in Japan, the section 'Land Laws, India' aims to create a tabular comparison between the two countries.

Land Trust Law, Japan

In Japan, trust business can only be carried out by entities licensed under the Trust Business Act (Act No. 154 of 2004, as amended) and financial institutions licensed under the Act for Financial Institutions' Trust Business (Act No. 43 of 1943, as amended). Until the early 1990s, trust business was monopolized by eight financial institutions, of which seven were known as trust banks. Following the reform of the financial system in recent years, today over 200 financial institutions and entities are licensed to carry out trust business in Japan (Kanda 2016).

Difference Between Commercial Banks and Trust Banks

Banks deal with deposit, lending affairs and other many relating businesses. On the other hand, trust banks in a limited sense receive trust of the money. They also receive trust of movable property and real property other than the trust of bonds and funds or money. In addition, they start taking a role in real estate brokerage or appraise when they are entrusted with the trust of real property. They may deal with inheritance affairs, too.

Indeed, from a functional perspective, there are four distinctive types of commercial trusts in Japan, and applying the basic principles recognized under the Trust Act to any of these types would lead to unwarranted results: (1) trusts similar to deposit taking and lending, for example, loan trusts; (2) trusts for asset management, for example, money trusts and securities' investment trusts; (3) trusts for securitization, for example, money claim trusts; and (4) trusts for business, for example, land trust (Kanda 2016).

Land Trust Structure

There are three bodies in trust; the entruster, entrustee and beneficiary.

When an entruster has an asset that he wishes to leave for the beneficiary, but he does not want to give it right away, he may entrust the entrustee with the asset with certain conditions for the beneficiary to receive the profit. This method can be used for trust by will (Fig. 9.1).

The entrustee must manage the trust asset by following the three rules stated below:

1. Due care of prudent manager

   The entrustee must manage the trust asset with care of prudent manager.
2. Duty of loyalty

   The entrustee must manage the trust asset for the beneficiary following the purpose of the trust. The entrustee must not use the trust asset for the benefit of himself or the third party.
3. Obligation to separately manage trust assets

The entrustee must manage the trust asset apart from the beneficiary's property or any other properties.

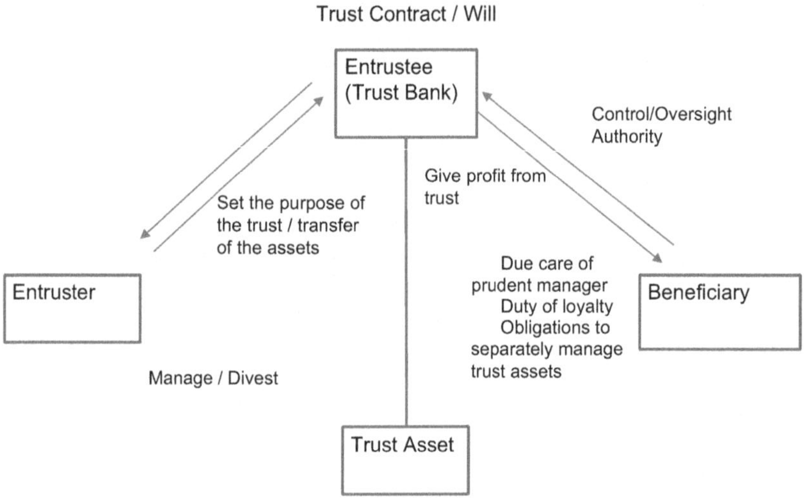

Fig. 9.1 Land trust structure

Trust of Real Estate

In cities like Tokyo, individuals who live in small houses may increase the utility of their land by consolidating the land and building apartments or office buildings on it. To do so, those individuals will need help from the trust bank or trust company. The land owners entrust their land to the trust bank, and trust bank builds a large building on the land to realize the effective utilization of the land. And those land owners can live in apartments within the building and receive the part of the profit as dividends from the trust bank. An individual land owner can gain more profit by this method.

Trust of Agricultural Land

After World War II, the Japanese Agricultural Land Act was changed, and ownership of agricultural land was shifted to tenant farmers from the major land owner. This realized the equality in agriculture in Japan. However, with the ageing of those farmers, there was no one to inherit the land or business. It was getting more challenging to use the land effectively.

Thus the idea of 'trust of agricultural land' was proposed. It posits that land owners entrust their agricultural land to trust banks and the trust banks manage the land. In this case, the trust bank aids young farmers who wish to farm the large consolidated land to use it and increase the utility. The land owners will receive the part of the profit as dividends. The consolidation of land leads to higher profits for land owners.

Presently, land owners are obligated to give the land to the trust bank. But in the future, giving only the usage right of the land to the trust bank can be an option, and in that case, registration of the land will be necessary. This way, land owners can maintain the ownership of the land and increase the profit by lending the land to the younger farmers through trust banks.

Another instance where the same method can be employed is in the case of ship owners or cowmen to increase the utility of their ships or cows. Individual farmers who own a small number of cows may consolidate the business with larger farmers by entrusting cows and that will help them to operate more effectively.

The method is to consolidate assets owned by individuals, entrust them to the trust bank and make a better use of the assets. It has a similar function to trust of money. Consolidating money to operate more effectively is the same as consolidating assets owned by individuals who are not able to maximize the utility of their assets by themselves or do not have the know-how; entrusting them to the trust bank can increase the utility of the assets. The same kind of method can also be used to revive local shopping streets with many closed small shops.

Land Laws, India

Property and law are born together, and die together. Before laws were made there was no property; take away law, and property ceases

—Bentham

India faces serious challenges in creating development processes that generate economic growth while being socially inclusive, ecologically sustainable and politically feasible and in accordance with the rule of law. Land is a State subject, but acquisition of property is on the Concurrent List, which means that both state assemblies and Parliament can make laws that govern it, but in case of conflict the central law will prevail.

Cities and towns are crucial to the economic well-being of India. For this, it is imperative that its cities and towns are transformed and pressures of new growth are dealt with so that they are more liveable, efficient and environmentally sustainable. Only then will the rapid pace of economic growth that India is undergoing be sustained and the targets of environmental sustainability of the world achieved. Land readjustment is advocated as an important land management tool in developing countries like India, where rapid urban growth is making the need for large investments in public infrastructure more evident.

The origins of the eminent domain law in India can be traced to the colonial state's need to create infrastructure to facilitate the movement of goods and people and enable commerce; these first 'public works' were typically canals and roads; later came railways, mines, and irrigation schemes; and even later came factories and other business establishments. One of the main reasons to use eminent domain, then and now, is to get the needed land cost effectively. Another, possibly even more important, reason is to get it quickly by avoiding protracted negotiations with numerous small landholders (including 'holdouts' or owners waiting for better offers) and sidestepping the legal problems of sorting out the considerable ambiguities about who 'owns' what. Therefore, eminent domain was and remains essential to collate multiple properties and own them 'free and clear' of legal encumbrances. The justification for the taking has always been 'public purpose', an all-encompassing term whose meaning and ambit have been debated in the courts from the very beginning (Table 9.1). It is also imperative to note, land is a State subject under the Indian Constitution, but acquisition of property is on the Concurrent List, which means that both state assemblies and Parliament can make laws that govern it, but in case of conflict the central law will prevail. This adds to the already existing complexities of land management in the country.

Conclusion

Many developing countries face a shortage of public funds to meet their huge infrastructure needs. In order to narrow the gap between investment needs and actual government allocations, private funds have to be injected into infrastructure investment. It is quite important to attract private investors into infrastructure by increasing the rate of return on infrastructure investment. The spill-over effects of infrastructure investment will increase revenue from corporate, income, sales and property taxes. The use of a land trust for infrastructure supplements this by creating a similar regulated method for better investment opportunities, along with creating a space for ensuring human security by creating more profit opportunities for the land owners. In the case of Bali, a highway was built over the ocean due to land owners not agreeing to give their ownership; this has led to an

Table 9.1 Comparing land law provisions between Japan and India

Indicators	*India*	*Japan*	*Recommendations*
Public purpose	According to Section 3 of the Land Acquisition Act, 1894, acquisition of land for 'public purposes' includes, among others: provision or planned development of village sites; provision of land for town or rural planning; the provision of land for planned development of land from public funds in pursuance of a scheme or policy of the government; and the provision of land for a corporation owned or controlled by the state.	Article 29 of the Japanese Constitution expresses the fundamental rights of Japanese citizens in regard to property, stating that '(1) the right to own property is inviolable; (2) property rights shall be defined by law, in conformity with the public welfare; and (3) private property may be taken for public use upon just Compensation therefor'.	The definition and scope of public purpose have been rather ambiguous and vague, leaving room for exploitation. A more concise definition will aid in creating a more comprehensive approach.
Trust	It does not exist within the Indian body, only for private parties with no state intervention under the Indian Trust Act, 1852.	The Japanese Trust Act, 2006, is a model legislation with scope for being implemented on the acquisition of agricultural land for infrastructure development.	The success of the Japanese Trust Act could be used to pursue changes in other Asian countries' laws.
Regulating authority	The absence of a one all-inclusive body as opposed to the existing framework of convoluted laws creates an unnecessary state of delay, confusion and lengthy procedure, which can and does get exploited for the benefit of one of the parties.	The Trustee, under Article 26, shall have the power to administer or dispose of property that belongs to the trust property.	The creation of the proposed trust bank will help in mitigating disputes and curtail lengthy legal procedures.
Determination of price	The government almost unilaterally decides on the price of compensation, which is in most cases contested by land owners.	Under Article 104 (1) of the Trust Act, the trustee and beneficiary mutually decide on price. This creates a more conducive environment, enabling the land owner to get a desired price, thus mitigating further disagreements.	The Japanese provision of allowing mutual price determination creates a more secure network of engagement between the land owners and the developers/buyers.

(*continued*)

Table 9.1 (continued)

Indicators	*India*	*Japan*	*Recommendations*
Damages and termination	The affected party is expected to seek legal assistance in case of damages and termination.	As per Article 164 (2), when a settlor and a beneficiary have terminated a trust at a time that is detrimental to the trustee, the settlor and the beneficiary shall compensate the trustee for any damages, as long as there is no compelling reason for the trust to be terminated at that time.	The strengthening of laws within the Indian framework with the proposed regulating body will lead to better dispute management.
Transparency	The Right to Information Act, 2005 and public interest litigations (PILs) are some mechanisms to verify if due process was followed.	The Trust Act offers transparency in matters of land compensation and determination of price.	The Indian system can be enabled with the inclusion of a regulating body, focusing only on land acquisition/land readjustment cases with similar powers as that provided to the land trust bank.

almost negligible, if any, spill-over effect from such a high expenditure-incurring project. Land trust is thus a way to mitigate obstacles that create disputes over land acquisition. The examples of India and Japan have strong potential to be applied to other Asian regions facing similar issues, which will contribute to the development of infrastructure in the region while sustaining the rate of return for land owners.

References

Agrawal, A., & Redford, K. (2009). Conservation and displacement: An overview. *Conservation and Society, 7*(1), 1–10.

Bhattarai, A. M. (2001). *Displacement and rehabilitation in Nepal: Law, policy and practice.* New Delhi: Anmol Publications Pvt. Ltd.

Guneratne, A. (2002). *Many tongues, one people: The making of Tharu identity in Nepal.* New York: Cornell University Press.

CHAPTER 10

Economic Development and Effects on Land

Shreyas P. Bharule

Introduction

Various studies in the past have tried to draw multiple parallels between the phenomenon of economic development and infrastructure investments. Studies have stratified the levels of economic development in the form of implementation location, scale and context with the location, but the presented arguments have mainly been indicative. The chapter focuses on a lesser highlighted area of the sequence of decisions which create externalities over the period that either helps in attaining the envisioned development or completely changes the form of spatial development. Such susceptibility of the uncertainty of a positive spill-over gives rise to specific intertemporal concerns. The purpose of the study is to shed some light on intertemporal concerns to resuscitate the incentive mechanism.

To explain the phenomenon of the association of spatial dynamics of the region with the decisions of the authorities, the chapter is divided into three sections: the first section is a critical review of concepts of economic development and its claimed links with infrastructure investments while

S. P. Bharule (✉)
The University of Tokyo, Tokyo, Japan

Asian Development Bank Institute (ADBI), Tokyo, Japan
e-mail: sbharule@adbi.org

N. Yoshino, S. Paul (eds.), *Land Acquisition in Asia*,
https://doi.org/10.1007/978-981-13-6455-6_10

focusing on theories of urban planning and economic development; the second section summarizes each of the road development plans that are implemented in India; and the third section focuses on the case of the Mumbai-Pune Expressway, giving special attention to the development around the expressway corridor.

The Interdependence of Road Development and Economic Development

Since the early twentieth century, establishing linkages between infrastructure development and economic development has been a common feature in economic analysis. One of the earliest and basic forms of infrastructure which has been extensively analysed is the development of roads. In the early twentieth century, Alonso (1964) argues that investment in roads provides increased accessibility to the land which in turn increases the land value resulting in economic growth. However, Black (2001) argues that it is difficult to prove that investment in highways spurts economic development. Regarding this, it has been extensively argued that transportation planning generally ignores the amelioration of transport infrastructure enhancements over land use especially during the plan and project evaluation.

Banister and Berechman (2001) discuss the connections between economic development and transport investments in developed countries. The analysis is stratified into national, regional and local levels. They concluded that location of the investment, the scale of the project and investment, as well as positive externalities are of equal importance. Not only are they important but if they are adequately backed by policy, political and institutional support, they can bring about positive spill-over in the form of economic development. However their approach was focused only on developed countries where economic development was already well augmented in the transport infrastructure development process. There were a certain set of limitations to the approach: firstly, with the differences in externalities, generalizing the findings of the investigation from developed to developing countries was not possible, and secondly, the explanation regarding the relationship between transport investment and economic development remains hazy in the light of other components: politics, policy and institutional support and so on. This makes the premise of establishing a transport infrastructure and economic development relationship questionable.

Gauthier (1970) highlights the three views about the relationship between economic development and transportation as positive, permissive and negative. The first view states that transport investment promotes economic growth and their hence their relationship is positive as transport facilities help in expanding production facilities. The second view states that transportation does not directly generate productive activities and elaborates that economic development is based on contextual conditions and is not a deterministic process hence being rather permissive. The third view completely dissociates the relationship calling it to be negative in nature. The view hints that over-investment in transport infrastructure has adverse effects on economic development due to relocation of economic activities from the urban core to peripheral areas of the city leading to change in the employment and income patterns.

Mohring (1976) pointed out that transportation improvements may change land values resulting in transfer of income in the population under the influence of the implemented infrastructure. This has been a function of the local dynamics also known as location externalities (see Martinez and Araya 2000). These arguments suggest that every transport infrastructure project must be treated differently in the context of various planning and economic conditions of the locations, as it might influence the spill-over by infrastructure investment.

Location Theories: Emphasizing the Existing Urban Hierarchy

The early urban planning and economic growth theories were focused on localism and contextual issues. Hoover (1936) assumed that there are no geographic changes in the cost of investment at different sources of production and suggested that the cost of investment may be assumed to remain constant across developed and developing countries. Based on this assumption they concluded that production costs are not related to geographic diversity and the ideal location is the point of least transportation cost.

Although Roy (2011), argues that as location theories have been very site specific, most of the planning models have a location-specific nature and vision. Confirming the policy side of the argument, March (2010) argues that the merit of a planning theory usually is dependent on the nature of the governance at the location. Hillier (2008) argues that a new theoretical base is required for spatial planning models, disentangling the complex challenges of today. This hints towards the idea that economic development impacts or rather the spill-over of infrastructure investments at a given location not only is confined to geography but involves several managerial and economic dimensions.

To summarize the current argument, Capello (2011) notes that location theories were unable to acknowledge the existence of several other activities by direct or indirect stakeholders, taking place at the same time. This hints towards the intertemporal concerns that might arise on the project location. Dichotomous location alternatives such as urban or rural areas or core or peripheral areas with varying local and contextual activities are also not included in the subject. This highlights the need to understand the location-based economic development in the context of the occurring urban system or hierarchy of centres which have led to the present locational issues.

Regional Growth Theories: Spatial Structure of a Region Affects Economic Growth

Regional growth is a long-term phenomenon. Regional growth can be attained incrementally with certain basic steps like having a holistic regional development plan for the area of influence. Hoover and Fisher (1949) suggested that a region undergoes five stages of economic developmental process. First, there is little investment or trade and populations are located according to the distribution of natural resources in the region. In the second stage there are certain improvements in transportation; thereafter the region starts developing some trade and local specialization. The third stage observes technological advancement in agricultural practices. The fourth stage is where the region is industrialized due to reduction in income through agricultural practices. In the final stage, a region specializes in tertiary industry and exports to comparatively less-developed regions. Isard (1951) confirms that the role of transportation costs remains critical in these five stages, he observed that reduction on transportation cost transforms a scattered production pattern into a more concentrated one.

However, North (1955) criticizes the regional economic growth theory mainly because of its lack of contextual consideration regarding the regions' level of development. As different industries have different transportation requirements, he suggests associating regional economic development studies with the economic history of the region. He stresses upon the establishments for exports and the competitiveness of the local economic system; however, to this, Amdam (2001) contradicts that this type of local economic setup will weaken over time if the urban development does not account for dependency on the region itself. Therefore, the investment in transport infrastructure with a different kind of regional structure may not bring similar type of spill-over development results.

Local Development: Context as the Driving Factor

Local development theories incorporate both exogenous and endogenous factors in enhancing the economic growth of a local area. Perroux (1955) suggests that economic growth is not uniform over a region but accumulates around poles in clusters. Regional development remains to be unbalanced because of agglomeration and the scale of the economy within these clusters.

Transportation is one of the key components in the process of agglomeration and clustering of the industries. Critics argue that confusion exists around the term 'growth pole' and needs to be reviewed in geographic and economic space. To this Rephann and Isserman (1994) add that the suburban areas will be the largest beneficiaries regarding economic development if they are close to an industrial growth pole of a large city.

During the 1970s, some studies undertaken in Italy showed that certain regions of the country were under rapid development, and this took place during the period of industrial crisis in Italy. The studies highlighted the presence of some endogenous factors in rapid development, as explained by Capello (2011). Soon after, several endogenous theories started focusing on localism, highlighting local contexts and trying to explain the phenomenon of rapid economic growth in the country.

The local development theories laid a greater emphasis on development than on economic growth. They highlighted the importance of context in the complex process of local development. Krugman (1991) proposed the 'New Economic Geography'; the new theory explains how historical incidents shape economic geography in which several hidden parameters can generate never-ending changes in spatial structures.

Scale, Context and Drivers

The location theories are very context sensitive compared to others. These theories, however, were not very successful at explaining local sensitives in an urban hierarchy. The regional growth theories could link the spatial aspect of economic growth to territorial population and economic distribution. These theories are predominantly context sensitive (see Capello 2011). The locational development theories primarily identified the determinants involved in locational development. All the theories have laid a greater emphasis on the complex development process involved in actual development than economic development over the long term. Although local development theories could study social issues, however, the theories remained fixed and determined with a dim focus on context.

Road Development Plans in India

The previous section made it very clear that the provisions for road transportation are one of the most vital for economic development of a region, for that matter a nation. Road infrastructure provides an opportunity for social integration in a safe and efficient manner while adding ease to the movement of goods and services. In India, multiple road development plans have been launched by the government since the early years of independence. The central idea of most of these plans caters to the increasing needs of the growing economy and demand in transport service. The following section summarizes the plans in their chronological order and highlights the National Highways Development Project (NHDP), the targets of which are envisioned to be achieved by the horizon year 2021.

A Chronology of Road Development Plans in India

Before independence of India in 1947, the Nagpur Road Plan was at work between 1943 and 1963. The road development plan classified the roads in India into national highways, state highways, major district roads, other district roads and village roads. Another important aspect of this first ever 20 Year Plan in India was that it designated the norms required for these categories mainly based on usages and carriageway. The target was to lay 200,000 km of surfaced roads where the farthest distance for developed areas and agricultural land was the target within 8 km, while around rural areas the maximum road grid size was to be within 16 km. Also, all villages on the census list were to be provided with road accessibility within the distance of 3.2 km. Although the plan was a 20-year vision, the target lengths were achieved in 1961 itself.

However, while the envisioned road lengths were achieved by the end the Nagpur Road Plan period, the other aspects of road frameworks were insufficient. The change in the country's financial, agricultural and industrial conditions of that period called for a revised plan and concept to further accelerate the growth of the country. Hence, the second 20 Year Plan was drafted and was known as the Bombay Road Plan (1961–1981). The plan envisioned to achieve 1 million km of total road length across the nation which will give an average road density of 32 km per 100 sq. km. The plan also includes the construction of 1600 km of expressways. The funds for highway financing should come out only from direct users but also from those who receive indirect benefits. Hence, there is a need for the introduction of spill-over mechanisms like betterment levy, land revenue schemes within the proximity of roads and additional taxation on motor vehicles using diesel oil. The plan also proposed the introduction of authorities to reduce roadside encroachments.

The third road development plan was called the Lucknow Road Plan 1981–2001. The vision aimed at constructing 1.2 million km of road length by 2001. The target was to bring all the villages in the country within 50 km of a national highway, ensuring that the net road density could be of 82 km per 100 sq. km. One of the goals contained in the plan was to construct expressways on corridors with dense traffic to provide speedy travel. All the aspects brought in the picture relevant factors such as the urgent need to connect the rural, hilly, tribal and backward areas with the nearest centres for administrative, health, market and educational facilities by building roads, thereby creating a positive spill-over. Due importance was given to energy and environmental conservation by improving the quality of roads and road safety measures. Consideration for the inclusion of non-motorized transport while designing roads was highlighted in the plan.

The current plan in place is called the Road Development Plan: Vision 2021. Proposed by the Indian Road Congress, the concerns taken by the vision are increasing the carrying capacity of highways, boosting the accessibility of the villages, encouraging private sector participation by incentivizing road development, and financing road-by-road tolls and mobilization of funds for boosting sectors associated with road infrastructure development. The National Highway Development Project has been split on to seven phases under which roads of various types have been classified (Refer Table 10.1).

Table 10.1 States of national highways under the NHDP (as on March 31, 2017)

Project	*Phase*	*Total length (km)*	*Already four or six lanes (km)*	*Under implementation (km)*	*Contracts awarded (km)*	*Balance length (km)*
National Highway Development Project (NHDP)	Golden Quadrilateral	5846	5846	0	–	–
	NS-EW Phase 1 and Phase 2	7142	6563	305	28	274
	Port connectivity	435	381	54	7	
	Phase 3	11,809	7507	2357	72	1945
	Phase 4	13,203	3773	6373	105	3057
	Phase 5	6500	2544	1424	9	2532
	Phase 6	1000	–	184	9	816
	Phase 7	700	22	94	4	584
	Total length	46,635	26,636	10,791	234	9208

Source: National Highway Authority of India Website: http://www.nhai.org/about-nhdp.htm

Development of National Highways During the 12th Five Year Plan (2012–2017)

The project is composed of the following phases:

- Phase 1 focuses on the Golden Quadrilateral, one of the most ambitious highway projects of length 5800 km; it connects the four metropolitan cities of Mumbai, Delhi, Kolkata and Chennai. It also connects other tier-two cities and towns between the metropolitan cities, connecting the urban hierarchy.
- Phase 2 focuses on the North-South and East-West Corridor; the total corridor length is of 7300 km and it connects the extreme points in the country. The North-South Corridor connects Srinagar (North) to Kanyakumari (South) and is 4,000 km long, while the East-West Corridor connects Silchar (East) to Porbandar (West) and is 3,300 km in length. This phase also includes connectivity to various ports and other aligned projects.
- Phase 3 mainly covers the areas not covered by Phases 1 and 2. The projects of existing national highway (NH) upgradation are covered under this phase. Since almost all the projects under this phase follow the Built, Operate and Transfer (BOT) basis, most of the private

players are involved in this phase. This phase includes connectivity to state capitals, implementation along high-traffic density corridors, tourist attractions and places of economic importance.

- Phase 4 converts the existing single-lane highways into two lanes. These are highways which are not a part of phases 1, 2 and 3.
- Phase 5 is still in progress as the government is yet to identify the stretches that will be most affected as a result of increase in road traffic over time. The plan envisions to upgrade the current four-lane highways to six lanes.
- Phase 6 focuses on constructing the expressways, connecting the major commercial and industrial cities.
- Phase 7 calls for improvements of city road networks by adding ring roads and bypasses to ease connectivity with national highways as well as helping in segregating intra-city traffic from the inter-city traffic passing on the highway. These improvements in turn help cities in proposing housing growth along the ring roads and bypasses. Although currently contributions to this phase are a part of city development plans where the cities propose ring roads according to their growth direction, the government is yet to identify an investment plan for this phase.

Intertemporal Concerns and Constraints During Road Development

Once a road segment is declared for implementation the immediate concerns are related to land acquisition, protests opposing the execution of projects and landowners filing petitions in the court to put the project on hold. Legal protests are in the form of petitions that demand a valid reason for landowners to contribute their properties and land for necessary land acquisition, while other protests take a strong form and result in damage to property. Many locals and landowners protest as highways pass through their areas, and the compensation demands vary from region to region. Some groups protest in the demand for underpasses and over-bridges while others protest fearing loss of their livelihood as their roadside may not receive an adequate number of highway users. Anti-social or political groups usually support most of these law and order issues.

Such intermittent hurdles cause excessive amount of delay in acquiring the land in some parts of the countries mainly due to formalities, petitions, stay order cases and the lack of state governments' full support to the central government's plans. Another area of regulation causing delays is the procedure of obtaining clearances from the Ministry of Railways and Ministry of Environment and Forestry (MoEF). Obtaining clearances involves coordination with several departments within the railways as well as the MoEF, which creates further delay. Road development also becomes a subject of political stir. These intertemporal issues create long-term effects on the project area of influence (Refer the section 'Case of Mumbai-Pune Expressway: Spill-Over by Intertemporal Concerns', Table 10.2). Shifting of urban utilities when implementing road projects in urban areas usually consumes maximum time. Urban physical infrastructure and utility

Table 10.2 Sequence of intertemporal events on the Mumbai-Pune Expressway project area

During	*Year*	*Decisions and activities that occurred in the same region*
Nagpur Road Plan 1943–1963	1950	The Indian government took up the task of developing the public sector
Bombay Road Plan 1961–1981	1966	Gadgil Committee Report was presented
	1967	Maharashtra Regional and Town Planning Act, 1966, was brought to effect
	1969	Banking sector nationalized—branches expanded in rural areas
	1970	Navi Mumbai was proposed in the Draft Regional Plan for Mumbai Metropolitan Region (MMR)
	1971	City and Industrial Development Corporation (CIDCO) registered under the Indian Companies Act, 1956. CIDCO was appointed as the city planning authority for Navi Mumbai
	1973	Vashi bridge over Thane Creek was opened for public
Lucknow Road Plan 1981–2001	1990	The Maharashtra state government commissioned a feasibility report for an expressway connecting Mumbai and Pune
	1992	Agricultural produce market was established in Vashi, the Mankhurd-Vashi Suburban Rail started
	1991	The 760-km-long Konkan Railway construction started
	1994	Mumbai-Pune Expressway Feasibility Report that was submitted was accepted and tenders for bidding were floated
	1997	Konkan Railway construction was completed
	1998	Project management contracts were appointed for all four sections
	May 1, 2000	Mumbai-Pune Expressway—Sections A and B were opened to public
	2001	Navi Mumbai become a million-plus city

(*continued*)

Table 10.2 (continued)

During	*Year*	*Decisions and activities that occurred in the same region*
Road Development Plan Vision: 2021 2001–2021	April 1, 2002	Mumbai-Pune Expressway was fully operational
	2008	Government of Maharashtra granted approval for Navi Mumbai International Airport
	2012	Navi Mumbai Airport Influence Notified Area (NAINA) was proposed by CIDCO
	2013	CIDCO was appointed as Special Planning Authority (SPA) of the NAINA
	2015	NAINA interim draft development plan was submitted by CIDCO
	2016	MSRDC appointed as SPA for villages along expressway. MSRDC SPA's boundary overlapped with NAINA and CIDCO boundaries
	2017	MSRDC appointed as SPA for 71 villages along expressway
	2018	Foxconn commits to invest US$5 billion along the expressway

lines like water, sewage, telecommunication and power supply lines need to be shifted with the directives of the concerned line departments of urban authorities, and it takes time depending on the capacity of the urban local body. Although these intertemporal issues are extremely difficult to avoid, they can be taken care of by establishing a land trust, a multi-institute collaborative land management institution as elaborated in Chap. 11.

To summarize, road development plans were predominantly focused on increasing the accessibility of various locations, with the underlying aim of enhancing the economic development of the country, assuming the spill-over of infrastructure investments. This mammoth task of road development was hence decided to be carried out in a planned manner. Although the envisioned targets for the lengths were achieved by almost each of the plans, various other externalities such as environmental, budget constraints, safety and amenities on the sideway were not emphasized. Currently, after private sector participation in infrastructure development, road infrastructure delivery has improved. Though, the envisioned economic development by the spill-over effect of the infrastructure development faces a lot of delays due to intertemporal issues of land acquisition, acquiring clearance from ministries, and involvement of local bodies for shifting of utilities which should be included in the planning process.

Case of Mumbai-Pune Expressway: Spill-Over by Intertemporal Concerns

In the last few years, the transportation sector was identified as one of the most crucial in the transmission of economic development. The Indian government has been successful in promoting privatization in physical infrastructure delivery. The issues of clustering of economic activities in a few major cities with detrimental effects on the development of peripheral regions were blamed on the imbalance in transport infrastructure demand and delivery (see Perroux 1955).

This section talks in detail about the transport infrastructure delivery system post-privatization. The case discussed in the following section is of the first expressway in India, connecting the city of Mumbai with Pune. Commonly, known as the Mumbai-Pune Expressway, it was later renamed as Yashwantrao Chavan Expressway (see Fig. 10.1). The expressway was a crucial piece of infrastructure for the government of Maharashtra as it helped in unclogging the bottlenecks of industrial development and production capacity along the corridor.

Fig. 10.1 Mumbai-Pune Expressway (Source: Photograph by author, 2016)

Since independence, India has been relying heavily on the public sector for economic development, funding its activities with budget allocations through national and state planning. Public sector companies were owned by the state and national governments. During the Lucknow Road Plan period (1981–2001), the state government of Maharashtra commissioned a feasibility study report for an expressway connecting Mumbai to Pune. The submitted report in 1994 recommended a ten-lane expressway to be built on the Built-Operate-Transfer (BOT) basis. These recommendations were accepted and required tender documents were prepared to invite the bids. Thus, starting a new policy to allow privatization in public infrastructure delivery represented a significant departure.

Though the project seemed very lucrative, the tender bidding process found few takers; the cost for securing clearances from ministries and avoiding hold-ups by protests, the cost of procurement of high-end equipment for construction, the uncertainty of receiving the returns from toll collection, unavailability of government subsidies and the unexpected decline of real estate values due to a fall in housing demand along the corridor were amongst the many factors that dissuaded private bidders.

Maharashtra governments example of indigenous privatization like the case of Konkan Railways corporation which floated competitive tenders' and private builders built about 760 km of railways within a period of 6 years (1991–1997) and financed the project through public bonds. Likewise, the Maharashtra State Road Development Corporation (MSRDC) was formed with a small equity of Rs five crores. The organizational framework of these companies was such that the management was answerable only to the government, which in turn made the role of the government significant in this model of indigenous privatization instead of awarding autonomy to private players.

For the development of the expressway, MSRDC acquired 646 ha of land for the right of way, 455 ha of land for quarry and dumping areas and 1338 ha for real estate development, which were expected to generate revenue in near future. The area for right of way was a 90-m-wide strip for the carriageway and for project support facilities like project site offices and storage of construction materials and equipment. The proposed design was envisioned as the first access-controlled expressway in India, with interchanges providing exits for other state and national highways for connecting traffic to regions other than Pune, over and underpasses facilitating villagers on either side of the carriageway.

The expressway that was envisioned to provide *economic development through speedy traffic*,[1] the slower traffic of large and heavy container trailers carrying goods from the southern part of India to the Jawaharlal Nehru Port in Uran, has provided larger amounts of spill-over along the carriageway. Transport service companies have benefited through reduction in travel time and distance with road infrastructure, despite the high toll they pay. Spatial developments observed along the expressway were the formation of multiple inland container depots (ICDs), under-occupied and incomplete high-rise residential towers, proposed industrial estates, proposals for resorts and other residential townships (see Figs. 10.2 and 10.3). The key reason for these fragmented developments was the absence of a collective development plan.

Despite having an egalitarian vision of development along the corridor, the supportive mechanisms of land development like development plans for the region, development control regulation and land regulation bodies like the municipal corporation were not in place. The permissible floor area ratio (FAR) in the villages around the expressway was low to attract

Fig. 10.2 Inland container depots near expressway (Source: Photograph by author, 2016)

Fig. 10.3 Residential high-rise towers being built (Source: Photograph by author, 2016)

any developer. In late 2001, Navi Mumbai became a million-plus city, attracting population and commercial activities from the Mumbai Metropolitan Region (MMR); this initiated new real estate development around the main and more accessible nodes in Navi Mumbai. Almost no major development apart from industrial warehousing and ICDs was seen along the corridor until 2012, a decade after the opening of the expressway.

After the sanction of Navi Mumbai International Airport in early 2008, real estate demand in the villages around the expressway picked up pace. Upon announcement, real estate development around the expressway interchanges also picked up. Although the surge in land prices and real estate transaction along the expressway was still relatively lower than what the government expected upon announcing the new airport, the lack of available infrastructure, lower levels accessibility and accessibility to the expressway were some of the major deterrents. A satellite city-aerotropolis development plan named Navi Mumbai Airport Influence Notified Area

(NAINA) was announced in 2012. The plan encompassed 270 villages and proposed to develop the region in three phases. All the villages, within the MMR through which the expressway passed, were now to be developed under the NAINA and by the City and Industrial Development Corporation (CIDCO). CIDCO was appointed as a Special Planning Authority (SPA) because of their immense success in planning and development of Navi Mumbai. CIDCO proposed new development regulations called the NAINA scheme; this allowed landowners to get an additional floor area ratio (FAR) and exercise development tools like the Transfer of Development Rights (TDR) against voluntary land pooling. After land pooling, the owner has to contribute 40% of their land to the development authority, of which 25% will be returned to the owner with amenities and the remaining 15% will be used for physical infrastructure.

The NAINA interim draft development plan proposed the development of 23 out of the 270 villages with the NAINA scheme too and was approved in 2015. Through the scheme, few townships were developed in the villages; the effects were evident, but they were more noticeable on villages and settlements closer to already-established growth centres, with some existing social and physical infrastructure. Although the Patalganga industrial estate, which was located outside the MMR and NAINA, had considerable impact on the real estate within the NAINA, investments by some of the leading industries like Cipla, Castrol and Idemitsu attracted working populations to settle closer to the Patalganga industrial estate (see Fig. 10.4), which resulted in a spike in real estate. The incoming investments of industrial giants resulted in demands for physical as well as social infrastructure, to which the NAINA scheme catered to significantly. Soon in late 2017, the government of Maharashtra appointed the MSRDC to become an SPA for 88 villages along the expressway corridor from the 270 under the NAINA scheme. Although the MSRDC has experience in expressway development, the planning experience has been negligible, though the authority has not proposed a development plan for villages to

Fig. 10.4 Patalganga industrial estate (Source: Photograph by author, 2016)

be developed. The authority has invited bids in 11 different locations for commercialization along the corridor on the land they had initially acquired during the implementation of the expressway. This initial attempt appears to be a seed to create growth centres along the expressway corridor and long-term attempt for economic development of the region.

Recently in early 2018, one of the global leaders in electronics, Foxconn industries, had announced to invest in the area managed under the MSRDC-SPA. This investment once complete might add to the economic development of the region.

Conclusions

The presented case of the Mumbai-Pune Expressway and NAINA region has multiple parallels with the literature presented in the earlier section. The arguments of location-specific planning decisions and clustering presented in the literature review have an almost perfect explanation to the presented case. Although through the case it can be stated that economic development is not just a function of development choices, investment and infrastructure provision but appropriate governance mechanisms and development regulations. These aspects of inclusion of externalities in development planning and implementation were practically absent, as can be seen through the various road development plans. Though the underlying vision of economic development of the serviced region was helping the plans to stay on target but only for the envisioned road lengths, the current road development vision intends to do the same but with the added categorization of the types of services to be allotted along the different types of phases. This attempt needs further collaboration with development authorities, to exchange their views on the possible developments along each phase and to come up with a vision.

In the case of NAINA before the implementation of the NAINA scheme, the villages were mere settlements; after implementation the same areas have started constructing four- to five-storeyed residential townships. However, it is not just the scheme which has contributed to the development but also the various externalities such as availability of jobs due to investments by industries, proximity to social and physical infrastructure, existing growth centres and enhanced accessibility to better goods and services. Until now the region has followed the five stages of economic development, and all throughout the examined case, the literature on growth pole development is reflected in the region. The termination of toll collection on the expressway is due in 2019. An incentive like this might increase the traffic volume along the corridor and may give rise to new developments around the already-established growth poles (see Isard 1951). Such contextual externalities have an added advantage in creating the 'new economic geography' in any region designed by contextual awareness.

Note

1. Dandekar, H. C., & Mahajan, S. (2001). MSRDC and Mumbai-Pune Expressway—A sustainable model for privatising construction of physical infrastructure? *Economic & Political Weekly, 36*(7). Retrieved from https://www.epw.in/journal/2001/07/special-articles/msrdc-and-mumbai-pune-expressway.html.

References

Alonso, W. (1964). The historic and the structural theories of urban form: Their implications for urban renewal. *Land Economics, 40*(2), 227–231.

Amdam, R. (2001). Empowering new regional political institutions: A Norwegian case. *Planning Theory & Practice, 2*(2), 169–185.

Banister, D., & Berechman, Y. (2001). Transport investments and the promotion of economic growth. *Journal of Transport Geography, 9*, 209–218.

Black, W. (2001). An unpopular essay on transportation. *Journal of Transport Geography, 9*(1), 1–11.

Capello, R. (2011). Location, regional growth and local development theories. *Aestimum, 58*, 1–25.

Gauthier, H. (1970). Geography, transportation, and regional growth. *Economic Geography, 46*, 612–619.

Hillier, J. (2008). Plan speaking: A multiplanner theory of spatial planning. *Planning Theory, 7*(1), 24–50.

Hoover, E. (1936). The measurement of industrial location. *The Review of Economic Statistics, 18*(4), 162–171.

Hoover, E. M., & Fisher, J. L. (1949). Research in regional economic growth. In *Problems in the study of economic growth* (pp. 173–250). Universities-National Bureau. Retrieved from https://www.nber.org/chapters/c9514.pdf.

Isard, W. (1951). Distance inputs and the space economy. *Quarterly Journal of Economics, LXV*, 188–198.

Krugman, P. (1991). *Geography and trade*. Cambridge, MA: MIT Press.

March, A. (2010). Practising theory: When theory affects urban planning. *Planning Theory, 9*(2), 108–125.

Martinez, F., & Araya, C. (2000). Transport and land-use benefits under locational externalities. *Environment and Planning, 32*(9), 1611–1624.
Mohring, H. (1976). *Transport economics.* Cambridge: Ballinger.
North, D. (1955). Location theory and regional economic growth. *The Journal of Political Economy, 63,* 243–258.
Perroux, F. (1955). Note sur la notion de pole de croissance. *Economic Appliqee.*
Rephann, T., & Isserman, A. (1994). New highways as economic development tools: An evaluation using quasi-experimental matching methods. *Regional Science and Urban Economics, 24*(6), 723–751.
Roy, A. (2011). Urbanisms, worlding practices and the theory of planning. *Planning Theory, 10*(1), 6–15.

CHAPTER 11

Land Pooling as a Means of Mitigating Land Displacement in India

Saloni Lakhia

Introduction

The 'land question' has invigorated agrarian studies and economic history since Marx and the early twentieth-century writers on agrarian questions. Various transformative 'moments' have inspired and stirred debates around land control: the spread of colonialism, the rise of nation-states and nationalisms, the invention and triumphalism of global markets, collectivizations and privatizations. The 'right' of sovereign on land has been a contested subject through history. Even in democracies, the exigencies of collective benefit versus individual land rights have been at loggerheads.

This chapter aims to further the policy framework as described in Chap. 9 (Yoshino, Paul et al.) by understanding the existing legal structure in India and exploring the feasibility of the land trust method in a nation as diverse as India. By doing so, the chapter aims to introduce the various stakeholders involved apart from the land seller and developer and how the existing legal system could be enabled to include a structure much like the land trust bank. Lam and Paul in Chap. 2 of this book have elucidated the benefits of using a multidisciplinary approach, and this chapter continues that through a more legal and policy lens.

S. Lakhia (✉)
Asian Development Bank Institute (ADBI), Tokyo, Japan
e-mail: slakhia@adbi.org

N. Yoshino, S. Paul (eds.), *Land Acquisition in Asia*,
https://doi.org/10.1007/978-981-13-6455-6_11

Over the past several years, the convergence of global crises in food, energy, finance and the environment has driven a dramatic revaluation of landownership. Powerful transnational and national economic actors from corporations to national governments and private equity funds have searched for 'empty' land often in distant countries that can serve as sites for fuel and food production in the event of future price spikes (Borras, Hall et al. 2011). With advancement in transportation and reduction in transport costs, large units can move to the hinterland, away from seaports and airports. This creates a potential conflict with local populations which have no role in such relocations of factors of production to their areas.

In countries that allow private landownership, compulsory land acquisition is the right and action of the government to take property not owned by it for public use. In the United States, this right is known as 'eminent domain', and the action is known as 'condemnation' (Eaton 1995). In Canada, the United Kingdom and Australia, the right and action are known as 'expropriation' (Boyce 1984), 'compulsory purchase' (Denyer-Green 2013) and 'compulsory acquisition or resumption' (Brown 1972), respectively. In each of these countries, compulsory acquisition of private property by the government is authorized by legislation.

Land Laws in Asia: Eminent Domain and Beyond

At the core of land acquisition law is the concept of eminent domain, which asserts the right of the state to acquire private property without going through a market transaction. The eminent Dutch jurist Hugo Grotius is credited with coining the term 'eminent domain'. By this conception all property belongs to the sovereign, and hence the state is at a liberty to do as it pleases with it. Effectively, it is well within the rights (legal domain) of the state to alienate and even extinguish such property or rights contained therein, with the rider that such an action is for public utility. The governing corollary here being the maxim that private interests are subordinate to greater public good. But it is added that when this is done the state is bound to make good the loss to those who lose their property.

In other words, in democratic market societies, eminent domain is not meant to be an unconstrained right of the state but one that should seek some balance between two ideas: one, that public good can be superior to private interest; and two, it is important to have the system of individual right to property and legal appeal when that right is violated. Therefore, if the right to private property exists (it cannot be taken for granted that it does because several national regimes have abolished or severely reduced this right), there should also exist some legal redress to eminent domain takings by the state. It is this balance, the tug of war, that again, at a fundamental level, is concerning. The origins of the eminent domain law in India can be traced to the colonial state's need to create infrastructure to facilitate the movement of goods and people and enable commerce; these first 'public works' were typically canals and roads; later came railways, mines and irrigation schemes, and even later came factories and other business establishments. One of the main reasons to use eminent domain, then and now, is to get the needed land cost effectively. Another, possibly even more important, reason is to get it quickly by avoiding protracted negotiations with numerous small landholders (including 'holdouts' or owners waiting for better offers) and sidestepping the legal problems of sorting out the considerable ambiguities about who 'owns' what. Therefore, eminent domain was and remains essential to collate multiple properties and to own them 'free and clear' of legal encumbrances. The justification for the taking has always been 'public purpose', an all-encompassing term whose meaning and ambit have been debated in the courts from the very beginning.

However, there seems to be a growing impetus for legal reforms coinciding with the land debate. Within Asia, land laws in general have gone through a transformation. Most of these legal reforms, pertain to better compensation, rights of non-title holders and the general safeguarding of those displaced, which were earlier ignored. Indonesia, through their law on Land Development for Acquisition Purposes, effective since 2014, reversed former decrees that ignored compensation for land users without formal titles. Vietnam through its new constitution and land law is also seeking to include those that will be displaced by development through identification and consultation before acquisition. Sri Lanka and Cambodia as well have sought for significant awareness and rigorous consultation within their legal provisions to safeguard the interests of the landowners as well as land users.

Background: Land Policy and Land Reforms in India

Agrarian relations have required a readjustment of ownership patterns of land. If we examine the history of India, though there are instances of considering land as a private property by individuals who had control over it, the practice of communities like that of tribals with *collective ownership of land* stands out. Land like many other gifts of nature was considered free for all by many communities who didn't bother to fix boundaries for private ownership. But colonial rule by the British saw a dramatic shift in the landownership patterns of India. Lands of many tribal/forest communities were seized by British cultivators and Zamindars, and land tax was widely collected through systems like Zamindari, Ryotwari or Mahalwari.

A rich-minority-landowning class and poor-landless-peasant class became symbols of Indian agrarian society. As the ownership of land for some reason or other stayed with the rich Zamindar class, they became more powerful year after year, accumulating wealth. The peasants, who actually cultivated land, were often in poverty and remained landless. When India adopted socialistic principles after independence, equality in all spheres—social, economic and political—was envisioned. Land reforms are essential steps towards social and economic equality as land is a fundamental asset needed for healthy development of an individual. As per the Indian constitution, land reform comes under the list of state subjects, and hence the responsibility for bringing up regulations for effecting land reforms lies with individual states.

Land policy in India has undergone broadly four phases since independence (Deshpande 2015)

1. The first and longest phase (1950–1972) consisted of land reforms that included three major efforts: abolition of the intermediaries, tenancy reform and the redistribution of land using land ceilings. The abolition of intermediaries was relatively successful, but tenancy reform and land ceilings met with less success.
2. The second phase (1972–1985) shifted attention to bringing uncultivated land under cultivation.
3. The third phase (1985–1995) increased attention towards water and soil conservation through the Watershed Development, Drought-Prone Area Development (DPAP) and Desert-Area Development Programmes (DADP). A central government Waste Land Development Agency was established to focus on wasteland and degraded land. Some of the land policy from this phase continued beyond its final year.
4. The fourth and current phase of policy (1995 onwards) centres on debates about the necessity to continue with land legislation and efforts to improve land revenue administration and, in particular, clarity in land records.

Land reform legislation in India consisted of four main categories: abolition of intermediaries who were rent collectors under the pre-independence land revenue system; tenancy regulation that attempts to improve the contractual terms faced by tenants, including crop shares and security of tenure; a ceiling on landholdings with a view to redistribute surplus land to the landless; and finally, attempts to consolidate disparate landholdings (Basu).

Landowners naturally resisted the implementation of these reforms by directly using their political clout and also by using various methods of evasion and coercion, which included registering their own land under names of different relatives to bypass the ceiling, and shuffling tenants around different plots of land, so that they would not acquire incumbency rights as stipulated in the tenancy law. The success of land reform has been driven by the political will of specific state administrations, the notable achievers being the left-wing administrations in Kerala and West Bengal.

Land Acquisition and Land Readjustment/Land Pooling

Land acquisition is a process in which the government acquires the land for the public purpose of the development of infrastructural facilities, industrialization and urbanization. The government provides compensation to the affected people and the facility of rehabilitation and resettlement. In the 1930s, British planners transferred the German idea of land readjustment (LR) to India; this scheme was then widely implemented in the state of Bombay, which was divided into the states of Maharashtra and Gujarat after India gained independence from Britain in 1947. The Bombay Town Planning Act was introduced by the British in 1915 and then replaced in 1954 by a 'new' Bombay Town Planning Act. Pursuant to the 1915 act, local authorities were enabled to prepare town-planning schemes to develop parts of the municipality.

Land Readjustment/Land Pooling

The term 'land readjustment' (LR) or 'land pooling' refers to a technique for managing and financing urban land development, whereby a group of neighbouring landowners on an urban fringe area are combined in a partnership or a government agency consolidates a selected group of land parcels for the unified planning, servicing and subdivision of land with the project costs being recovered by the sale of some of the plots for cost recovery and the distribution of the remaining plots back to the landowners to develop or to sell for development (Archer 1992, 1994).

In this method, a public authority, such as a planning agency or a development board, brings together farmers/landowners under a scheme. This is done under legislative sanction of a town/urban planning act. This does not involve acquisition of land, and hence eschews the complications of transfer of ownership and attendant issues regarding compensation, etc. The land is pooled, with the consent of the stakeholders, and a master developmental plan is constituted, for providing infrastructural, logistical, power and other civic needs. A betterment charge is levied for these activities. The final plot allotted is proportional to the original holding, preferably near to its original location.

In Germany, LR was intensively employed in the postwar reconstruction of damaged cities and the accommodation of the recent wave of urbanization (Doebele 1982). Similarly, LR is the key part of the urban planning system in Japan. Since 1954 when the Land Readjustment Act was put into effect in Japan, LR was used for the development of new cities, prevention of disorderly growth and urban planning.

In Spain, although the practical experience of LR was unsatisfactory until the mid-1990s, after the legal reforms with the Valencia Regional Planning Law of 1994, LR (and if necessary, compulsory LR) became the standard procedure. Since then, LR has been implemented all around the Valencia Region as well as other Spanish regions in hundreds of cases, involving thousands of hectares. In addition, almost all the major real estate developments in Spain are performed using LR (Gielen 2016).

The process of LR in a nutshell involves the pooling and reconfiguring (re-parcelling) of underperforming contiguous land parcels and, then, installing urban infrastructure there. As an end result, the process facilitates coordinated and planned development where all landlocked parcels get access to planned roads and infrastructure. LR can also be characterized as a process by which private lot areas are reduced in size to create public land. The carved public land is then used for widening or, straightening out streets, developing public amenities and finally, creating developable lots and blocks (Mittal 2014) (Table 11.1).

Table 11.1 Land readjustment (LR) methods applied in different countries according to their characteristics

Key characteristics	*Japan*	*Germany*	*France*	*Sweden*	*Australia*	*South Korea*	*Turkey*	*India*	*Taiwan*	*Indonesia*	*Nepal*
Legal structure	LR Act of 1982	Baugesetzbuch (1987)	Article L 322-1-10 in the Code de l'Urbanisme (1985)	A Joint Development Act (1987)	Sections 6, 7 and 13 of the Town Planning Development Act 1928–1996	LR Act (1966)	18th article of Development Act (1985)	Maharashtra (1996 Regional and Town Planning Act) Gujarat (1976 Town Planning and Urban Development Act)	Articles 56, 76 and 161 of bylaws to Equalization of Urban Land Rights Law (1957)	No law to authorize and regulate LR	Development Committee Act 2013 (1954) The Town Development Act 2045 Para. 12, Land Reform Act 2021 and Bylaw 2060
Sector undertaking LR projects	Individuals associations Local governments (municipal, prefectural) Administrative agencies Public corporations	Local governments (always) Landowners Developers	Individual landowners A group of landowners	Landowners associations	Local governments (municipalities) Associations	Private landowners Associations of landowners Municipal or provincial governments Ministry of Construction	Municipality special Provincial administration	Local governments	Local governments Landowners	Municipalities	Committee of landowners Municipalities
Participation landowners to LR projects	Compulsory for LR with public initiative At least two-thirds of both landowners and lessees (by number and area) must consent to an application for private initiative	Compulsory for LR with public initiative	Voluntary participation (the support of at least two-thirds of landowners or, in some conditions, at least half of landowners)	Voluntary participation	Compulsory for LR with public initiative At least two-thirds of both landowners and lessees (by number and area) must consent to an application	Compulsory for LR with public initiative At least two-thirds of both landowners and lessees (by number and area) must consent to an application	Compulsory	Voluntary participation	At least one-half of landowners (by number and area) must consent to an application	Landowner consent (100 per cent landowner agreement)	Majority of landowners' consent

(*continued*)

Table 11.1 (continued)

Key characteristics	*Japan*	*Germany*	*France*	*Sweden*	*Australia*	*South Korea*	*Turkey*	*India*	*Taiwan*	*Indonesia*	*Nepal*
Cost recovery	Land contribution for public spaceCost-equivalent land for cost of the project	Land contribution for public spaceCost-equivalent land for cost of the project	Land contribution for public use Cost-equivalent land for cost of the project	Land contribution for public use Cost-contribution within area	Land contribution for public space Cost-equivalent land for cost of the project	Land contribution for public space Cost-equivalent land for cost of the project Land contribution can be made for low-cost housing	Land contribution for public space	Land contribution for public space	Land contribution for public space Cost-equivalent land for cost of the project	Land contribution for public use	Land contribution for public space Cost-equivalent land for cost of the project
Amount of land contribution	Land deduction rate is not determined (usually 20 per cent deduction for communal land requirements and 10 per cent set aside of cost-equivalent land)	No more than 30 per cent of market value of the land (in value basis) No more than 30 per cent of area of the land (in area basis)	Land deduction rate is not determined	Land deduction rate is not determined	Land deduction rate is not determined	Land deduction rate is not determined (usually 24 per cent to 28 per cent deduction for communal land requirements and 8 per cent to 10 per cent set aside of cost-equivalent land)	Maximum contribution in LR projects is 40 per cent If the contribution percentage within the project area is more than 40 per cent, the difference must be expropriated by the municipality	Up to half of increment in value	The proportion of land is not more than 40 per cent for public purposes and for cost-equivalent land (in Kaohsiung)	20 per cent of previous land holding of landowners	Land deduction rate is not determined Contribution share depends on the total cost to be shared by all the landowners

(*continued*)

Table 11.1 (continued)

Key characteristics	*Japan*	*Germany*	*France*	*Sweden*	*Australia*	*South Korea*	*Turkey*	*India*	*Taiwan*	*Indonesia*	*Nepal*
Assistance and support of the public	National and prefectural subsidy Low, zero-interest loan	All procedural costs	State awards grants at the initial stage	Preparatory investigation loans at the initial stage	Initial costs are covered by local governments, but these costs must be charged to the project	Deficits are covered from general city budget in public initiative projects	All procedural costs	All extra costs (outside of half of the increment value) are borne by local authority	-	All procedural costs (outside of land contribution) are borne by local authority	Assistance and support of public is nonexistent
The way of distribution stage	Value basis Area basis	Value basis Area basis	Value basis	Value basis	Value basis	Value basis Area basis	Area basis	Value basis	Value basis Area basis	Area basis	Value basis
The determination of valuation	Valuation by street value Amount of area for area basis	Market value (the values of pre-readjustment and post-readjustment)	A value fix by assembly of landowners before and after LR (value after LR represents initial plus increment)	The probable market value of before and after LR	Market value of land	Valuation by street value Amount of area for area basis	Amount of area for area basis	Arbitrary In the process, original plot value, semi-plot value and final plot value are determined; final plot value is found by a simple formula, but this unit price is adjusted considering some parameters	Pre-adjustment value (location, access, siting, sunlight, development feasibility)	Amount of area for area basis	Valuation by street value
Adjustments after LR	Money payments	Money payments	Money payments	Money payments	Money payments	Money payments	In the form of area	Money payments	Money payments	–	Money payments

(continued)

Table 11.1 (continued)

Key characteristics	*Japan*	*Germany*	*France*	*Sweden*	*Australia*	*South Korea*	*Turkey*	*India*	*Taiwan*	*Indonesia*	*Nepal*
Inclusion of infrastructure constructions and costs in LR process	Included	Not included	Included	Included However, municipality takes responsibility for general infrastructure	Included	Included	Not included	Included	Included	Not included	Included
Size criterion of LR project area	At least 5 ha	Nonexistent	Nonexistent	Nonexistent	Nonexistent	Nonexistent	LR projects are to be designed for areas with sizes equal to or larger than the size of a single residential block	100 ha (under the Gujarat Town Planning and Urban Development Act)	Nonexistent	Nonexistent	Nonexistent
Intervention to ownership	–	–	–	–	–	–	Although co-ownership is not converted into sole ownership through LR, there is an exception for some conditions in the law	–	Co-ownership is not converted into sole ownership through LR	–	–

Source: Turk (2008)

Land Acquisition Laws, India

The old land acquisition act had been in place since 1894, passed by the British government, and it was replaced by the new act on 26 September 2013 after seven to eight years of consultations. The 1894 Land Acquisition Act lacked consent for approval which allowed forced acquisitions, no real grievance redressals, no safeguards, absolute disregard to resettlement and rehabilitation of the displaced and low rates of compensation. Research into displacement in India showed estimations of around 65 million internally displaced people (IDPs) by development projects in the period between 1950 and 2005. Only 33 per cent of the IDPs were resettled in a well-planned manner. Some consequences of this include armed clashes, violent opposition and political instability. These various problems made the need for a new, more comprehensive act particularly constant. The new land act of 2013 Land Acquisition, Rehabilitation and Resettlement Act (LARR) aimed to take these issues into account and provided for acquisition for 'public purpose' only, in which public purpose was defined more specifically than before. Also, the urgency clause was made more stringent, only applicable now in cases of national defence and security or following natural calamities. The acquisition of multi-cropped land is only allowed under exceptional circumstances. In the case of government-led land acquisition, the approval of 70 per cent of landowners is required; in the case of acquisition by private entities, 80 per cent is needed. In addition to landowners, livelihood losers (including labourers, tenants and sharecroppers) can also now claim compensation (Sanyal 2013). These elaborations are in addition accompanied by an extensive package of resettlement and rehabilitation requirements, including a long-term subsistence allowance and a one-time off resettlement allowances. The LARR is controversial, both among industry representatives and among social activists. Some main deficiencies include the following: prices are based on average sale prices in the vicinity, which are often disgustingly understated and regularly represent distress sales; huge power and information unevenness are present between buyers and sellers; industry feels that the cost of acquisition will rise to impractical levels in the available land; and large delays are present in the acquisition process due to the required social impact assessments. So the land acquisition policies in India should be framed in an efficient manner in favour of both the affected landowners, and government restrictions on multi-cropped land may lead to major reduction.

Stakeholders Involved

Local Governments

India is a federal republic with three spheres of government: central (union), state and local. The 73rd and 74th constitutional amendments give recognition and protection to local governments and in addition each state has its own local government legislation. By empowering this local government, an equivalent of the land trust bank as discussed in Chap. 9 by Yoshino, Paul et al. may be feasible for implementation at a local level. By involving the local actors, here, the local government, the land pooled can be transferred for use as and when needed as per the land trust bank initiative, with a lower dissent probability amongst the landowners and users. The local government is in one of the best positions to better facilitate a smooth land transfer from the users to the developers.

Agricultural Co-operatives/Societies

In India, there are four major types of Cooperatives: (i) The Primary Agricultural Credit or service societies, (ii) Agricultural non-credit societies, (iii) Agricultural cooperative marketing societies and (iv) Co-operative Farming Societies. Though they initially gave only credit, post Independence, in states like Maharashtra and Gujarat, these cooperatives have been major engines of socio-economic development. They were an integral part of the Five Year Plans. They assist in various aspects like production, processing, farming and marketing-stopping the cycle of indebtedness and helping them break away from the clutches of moneylenders. Since Cooperatives are well established and have deep ties in communities they cater to, it would be imperative to involve them in planning infrastructure development.

Sectoral Banks

The National Bank for Agriculture and Rural Development (NABARD) is as an apex financing agency for the institutions providing investment and production credit for promoting the various developmental activities in rural areas. It coordinates the rural financing activities of all institutions engaged in developmental work at the field level and liaisons with Government of India, State Governments, Reserve Bank of India (RBI) and other national-level institutions concerned. Furthering the land trust model, it would be beneficial to involve NABARD in the land pooling functioning, empowering it with roles similar to the Land Trust Bank (See Chap. 9).

NGOs and Local Groups

After the institutions, it is the role of civil society that plays an integral role in a peaceful transfer. In the case of Maharashtra and the *Samruddhi Mahamarg* (Prosperity Highway) Project, the government faced multiple protests, which portrayed the unification and negotiation power a mobilized civil society could play. The land trust bank can involve members of civil society as third-party board facilitators who would be able to consider the rehabilitation and resettlement issues faced after the completion of land transfer.

Conclusion

Although there are difficulties in project areas due to landowners withholding their land from sale (including farmers, developers, land speculators and investors), many landowners can be encouraged to participate in land pooling projects when there is a possibility of their land gaining a significant increase in market value (Archer 1992). In terms of cost recovery, land pooling can increase the efficiency of urbanization at a reduced cost since the project site and infrastructure rights of way do not have to be bought or compulsorily acquired. The cost of infrastructure works and subdivisions can be financed with a short-term loan and then quickly recovered by the sale of some of the new building plots. Using it in land assembly, infrastructure and development costs can be substantially recovered from within the project. Differently from the common land-assembling methods, land pooling has the potential to overcome the holdout and free-rider problems of land management strategies. Moreover, using LR, it is possible not only to recover the cost of installing a complete infrastructure but also to capture the additional socially created value that can be used to subsidize low-cost housing or, indeed, for any public purpose. As stated by Vitikainen (2004), the Land pooling procedure is justified not only based on the involved costs and the efficiency of the method but also based on its fair treatment of landowners, improvements in plan quality, savings to the community and environmental benefits. Furthermore, it facilitates the participation of property owners in the process, ensures a fair distribution of development costs and profits created by spatial plans (Sonnenberg 1996) and preserves the original ownership structure and social networks. Although it theoretically provides better land management, in reality, only a few countries achieve positive outcomes. In the remaining countries, the procedure is still not introduced or the usage and success levels are far behind expectations. Therefore, the land pooling or land readjustment systems that are not successful or not accepted as the main land management and land assembly tool by the countries should be evaluated to clarify the problems that need to be solved and define the performance gaps that need to be addressed. To this end, countries should test their existing systems and compare the results with the best or expected results of an ideal land pooling system to identify the problems in their strategies and the performance gaps in their models/systems that need improvements (Mittal 2014). By understanding how land pooling can be efficiently implemented and maintained, it is possible to define the good practices and the success factors in terms of different aspects that should be addressed when the method is being introduced to a country for the first time or when existing land management policies are being improved.

References

Archer, R. W. (1992). Introducing the urban land pooling/readjustment technique into Thailand to improve urban development and land supply. *Public Administration and Development, 12*, 155–174.

Archer, R. W. (1994). Urban land consolidation for metropolitan Jakarta expansion, 1990–2010. *Habitat International, 18*(4), 37–52.

Borras, S. M., Jr., Hall, R., Scoones, I., White, B., & Wolford, W. (2011). Towards a better understanding of global land grabbing: An editorial introduction. *The Journal of Peasant Studies, 38*(2), 209–216.

Boyce, D. E. (1984). *Network models in transportation land use planning*. Elsevier Science Publishers BV.

Brown, D. (1972). *Land acquisition: An examination of the principles of law governing the compulsory acquisition or resumption of land in Australia and New Zealand*. Butterworths.

Denyer-Green, B. (2013). Compulsory purchase and compensation. *Estates Gazette*.

Deshpande, A., Patel, D., Patel, D., Jadeja, J., & Desai, K. (2015). Change in the land use pattern and water bodies of Vadodara city, India with respect to securing severe floods. *Journal of Environmental Research and Development, 10*(1), 175–186.

Doebele, W. A. (Ed.). (1982). *Land readjustment: A different approach to financing urbanization*. Lexington, MA: Lexington Books.

Eaton, A. (1995). Impact of urban renewal or land development initiatives on African-American neighborhoods in Dade County Florida. *Howard Scroll: The Social Justice Law Review, 3*, 49.

Gielen, D. M. (2016). Proposal of land readjustment for the Netherlands: An analysis of its effectiveness from an international perspective. *Cities, 53*, 78–86.

Mittal, J. (2014). Self-financing land and urban development via land readjustment and value capture. *Habitat International, 44*, 314–323.

Sanyal, J. (2013, April 9). The Land Acquisition, Rehabilitation and Resettlement Bill 2011—A jurisprudential critique. Retrieved from SSRN: https://ssrn.com/abstract=2294079 or https://doi.org/10.2139/ssrn.2294079.

Sonnenberg, J. (1996). The European dimensions and land management-policy issues (land readjustment and land consolidation as tools for development). In *Workshop on Land Management in the Process of Transition*. FIG Commission, 7.

Turk, S. S. (2008). An examination for efficient applicability of the land readjustment method at the international context. *Journal of Planning Literature, 22*(3), 229–242.

Vitikainen, A. (2004). An overview of land consolidation in Europe. *Nordic Journal of Surveying and Real Estate Research, 1*(1), 25–43.

CHAPTER 12

Conclusion

Naoyuki Yoshino and Saumik Paul

The acquisition, administration, and allotment of land for the purposes of development are intimately tied up with the issues of rights, local aspirations, power structures, and economic factors. This book has aimed to explore through diverse perspectives of the land grab crisis and its impact on those displaced within Asia. Throughout we have sought to establish that land is a crucial factor in attaining sustainable development and approaching it from just an economic perspective would mean to isolate the multiple other factors that involve land management. Evidence from our regional case studies also pointed for the need of a policy framework that we proposed as the land trust method of Japan, explained in Chap. 9.

However, there are points of concern to this method. For instance, the process is followed strictly only in Japan. That leads to a restricted case study analysis, as Japan is one of the leading developed countries within Asia, if not globally, and the existence of well-established institutions facilitates the smooth functioning of land trust method. Developing regions lack this institutional framework completely, or in some instances like the Philippines and Indonesia, they aren't as competent. Another critical factor is the diversity, not just regarding social systems, which we looked at in

N. Yoshino • S. Paul (✉)
Asian Development Bank Institute (ADBI), Tokyo, Japan
e-mail: nyoshino@adbi.org; spaul@adbi.org

N. Yoshino, S. Paul (eds.), *Land Acquisition in Asia*,
https://doi.org/10.1007/978-981-13-6455-6_12

Chaps. 5 and 7, but also concerning the very understanding of land and its relations with the people in a region. The legal and other binding structures lack the uniformity shared by the Western half, thus making it challenging to fit land trust in a practical and institutional framework.

Moreover, there are external factors which need addressing. For example, if one were to consider the resettlement cost, in Asia, it would be pertinent to raise the very availability of land for the very purpose. The global fluctuating markets make it challenging to estimate compensation, and an organized spillover is unfeasible without strong governance and local body structures.

Despite the above, there are still advantages of employing the method, which make it worth exploring further. For one, by just proposing the counter-factual and studying the failure of the existing systems, it is worth claiming the feasibility of land trust method. By identifying stakeholders and alternates, we have tried to bridge the gap of adaptability in diverse Asian regions.

Issues worth noting for further research may include but are not limited to the following: securing land tenure to achieving SDGs, studying the process of allocation of land for redevelopment, securing rights of indigenous communities, looking at tangible and intangible losses, and re-thinking the politicization of land development.

We conclude this book by reiterating that land is not purely "economic" or "social" in nature but extremely diverse. In the present global order, the SDGs dictate much of the development space, and thus it may be wise to integrate them into future discourses. For example, land use influences the environment or biodiversity (goal 13) just as much gender equality (goal 5) influences land access. Thus, by furthering the land debate, we may be able to achieve more than those SDGs pertaining to economic growth (goal 8), sustainable cities (goal 11), and responsible production (goal 12).

The manufacturer's authorised representative in the EU is Springer Nature Customer Service Centre GmbH, Europaplatz 3, 69115 Heidelberg, Germany. If you have any concerns regarding our products, please contact ProductSafety@springernature.com

Printed and bound by CPI Group (UK) Ltd, Croydon, CR0 4YY
30/09/2026
02225439-0001